Film Ecology

Using the Regenerative economic model – also known as Doughnut Economics – Susan Hayward offers a thought-provoking sketch for a renewed, tentatively revolutionary approach to both film theory and film practice.

This book attempts to answer the questions posed by T.J. Demos (in *Against the Anthropocene*, 2017): how do we find a way to address planetary harm and the issues it raises within the field of Film Studies? How do we construct a theoretical model that allows us to visualise the ecological transgressions brought about by the growth-model of capitalism which is heavily endorsed by mainstream narrative cinema? By turning to the model set out in Kate Raworth's book *Doughnut Economics* (2017) and adapting its fundamental principles to a study of narrative cinema, *Film Ecology* proposes to show how, by using this model, we can usefully plot and investigate films according to criteria that are not genre/star/auteur-led, nor indeed embedded in anthropocentric theoretical models, but principles which are ecologically based. These arguments are brought to life with examples from mainstream narrative films such as *The Giant* (1956), *Mildred Pierce* (1945), *Erin Brockovich* (2000), *Wall Street* (1987), *Hotel Rwanda* (2004), and *Missing Figures* (2016).

This approach will inspire film practitioners, film theorists, critics and analysts, film students, and film lovers alike to consider how they might integrate this Doughnut model into their thinking or work as part of their process.

Susan Hayward is Emeritus Professor of Cinema Studies at Exeter University. She is the author of numerous books on French Cinema and *Cinema Studies: The Key Concepts* (now in its fifth edition).

Film Ecology
Defending the Biosphere — Doughnut Economics and Film Theory and Practice

Susan Hayward

Routledge
Taylor & Francis Group
LONDON AND NEW YORK

First published 2020 by Routledge

2 Park Square, Milton Park, Abingdon, Oxon OX14 4RN
605 Third Avenue, New York, NY 10017

Routledge is an imprint of the Taylor & Francis Group, an informa business

First issued in paperback 2021

Copyright © 2020 Susan Hayward

The right of Susan Hayward to be identified as author of this work has been asserted by them in accordance with sections 77 and 78 of the Copyright, Designs and Patents Act 1988.

All rights reserved. No part of this book may be reprinted or reproduced or utilised in any form or by any electronic, mechanical, or other means, now known or hereafter invented, including photocopying and recording, or in any information storage or retrieval system, without permission in writing from the publishers.

Notice:
Product or corporate names may be trademarks or registered trademarks, and are used only for identification and explanation without intent to infringe.

Publisher's Note
The publisher has gone to great lengths to ensure the quality of this reprint but points out that some imperfections in the original copies may be apparent.

British Library Cataloguing-in-Publication Data
A catalogue record for this book is available from the British Library

Library of Congress Cataloging-in-Publication Data
Names: Hayward, Susan, 1945– author.
Title: Film ecology: defending the biosphere: Doughnut economics and film theory and practice / Susan Hayward.
Description: London; New York: Routledge, 2020. | Includes bibliographical references and index.
Identifiers: LCCN 2019058538 (print) | LCCN 2019058539 (ebook) | ISBN 9780367265519 (hardback) | ISBN 9780429293801 (ebook)
Subjects: LCSH: Environmental protection and motion pictures. | Environmentalism in motion pictures. | Ecology in motion pictures. | Capitalism in motion pictures. | Motion pictures—Philosophy. | Ecocriticism.
Classification: LCC PN1995.9.E78 H39 2020 (print) | LCC PN1995.9.E78 (ebook) | DDC 791.43/6553—dc23
LC record available at https://lccn.loc.gov/2019058538
LC ebook record available at https://lccn.loc.gov/2019058539

ISBN: 978-0-367-26551-9 (hbk)
ISBN: 978-1-03-217297-2 (pbk)
DOI: 10.4324/9780429293801

Typeset in Times New Roman
by codeMantra

For Sandra Cook
A great friend and supreme conversationalist
Whose estimable knowledge of film and theatre has
kept us in Dialogue for some 30 years.

'Only humans have the power to alter the ecosphere irrevocably and therefore only humans have the responsibility to manage it.'
(Regenia Gagnier, 2018, 226)

Contents

Introduction   1

1   Film and ecology, Doughnut Economics, and film theory   4

2   Film and the Anthropocene: dirty Capitalism – *Mildred Pierce* (1945), *Tulsa* (1949), and *Giant* (1956)   17

3   Where we are now and where we need to be: from source and sink to take→make→use→re-use – *Erin Brockovich* (2000), *Wall Street* (1987), and *The Wolf of Wall Street* (2013)   46

4   A moral imperative to REVOLT: *Hotel Rwanda* (2004) and *Caphernaüm* (2018)   76

Bibliography   113
Index   115

Introduction

When I started thinking about this project just over two years ago, I was full of despondency and gloom that no one (amongst the higher-ups) was really listening to the resounding voices of scientists, ecologists, and all manner of ecocide campaigners worldwide who were saying that CLIMATE CHANGE is for real, it's happening and if we do not take steps right now to bind ourselves to the Paris Agreement (at the very least), to commit to good ecological practice and law in all areas, we are inviting our own extinction within less than 100 years. The then newly elected President Trump of the USA had just ripped up America's signed copy of the Paris Agreement; other countries naturally felt the belt (or noose) had loosened around their obligations and so continued to move at a snail's pace to meet any CO_2 targets. However, a voice in the back of my head reminded me that in every negative there is some perceivable positive. And, presently, two years on, this would appear to be the case. There is some room for optimism. President Trump's continued Climate Change Denialism has, to a degree, back-fired – at least in countries outside the USA. People of all ages, especially the young (thank goodness), have started to bang their drums and have a presence both in Parliament and at the United Nations where their voices are treated with respect. In the arts, the Anthropocene and its effects are noticeably at the heart of numerous plays, films, television dramas, nature programmes and documentaries, painting, sculpture, novels, music, and poetry. School children are demanding the right to have a decent environment in which to live (surely a planetary and human right). The effects of plastic pollution are now an identifiable reality not just in the oceans (where the levels are truly shocking) but also within everything we digest and therefore within us. So too is the critical effect of air pollution particles – caused by vehicles and industry – on the world's population.

Thanks to the pressure we have put on the planet, we are responsible for the first geological epoch to be shaped by humanity. The effect of this planetary plundering is such that in just two centuries we have moved from the Holocene (a biosphere in which humanity and the planet were in balance, and in which we have survived well for 12,000 years) to the Anthropocene in which humanity and the planet are out of kilter, thanks to human (Anthros) technology. Meantime, we have failed to take responsibility for the planetary harm we do, to the effects of the harm that this causes to other peoples forced into migration because the land they lived in can no longer provide for their essential needs – as indeed is documented in Gianfranco Rosi's *Fuocamare/Fire at Sea* (2016), Ai Weiwei's *Human Flow* and Vanessa Redgrave's *Sea Sorrow* (both 2017).[1]

Two years ago, it was a case of where to begin. Increasingly, it has become a case for clarity. Two years ago, I was asked by BAFTSS to give a keynote talk at their 2018 annual conference, which was entitled Revolutions, at the same time I was reading the economist Kate Raworth's wonderful book *Doughnut Economics, Seven Ways to Think Like a 21-st Century Economist* (2017), in which she provides us with a remarkable model for regenerative economics, establishing a clear model by which to understand why growth economics do not (ultimately) work, but circular economies do. In her book, Raworth shows how circular economies can function and, in turn, bring the needs of humanity back into balance with the means of the planet. It seemed to me that this was where my paper on Revolutions needed to go. That exploration is brought to you now in this extended study in the hopes that ideas here will inspire film students, practitioners, and theorists to see how their work can benefit from such a model. The intention is never exclusive – for there are many models to be drawn upon, of course. Put simply, it is my hope that this model opens further new doors of thinking.

To the question, how do we find a way address this planetary harm and the crucial issues it raises within the field of Film Studies, my initial answer is by investigating two distinct economic models. The first is one that already exists, the Capitalist growth model which is heavily endorsed by much of mainstream narrative cinema – and we shall take a look at the implications of this model in Chapters 1 and 2.

---

1 As a measure of the perceived importance of these documentaries, Rosi's film received the Golden Bear at the 2016 Berlin Film Festival. Ai Weiwei's film was nominated, in 2018, for an Oscar. Redgrave's film was selected for the Special Screenings Section at the 2017 Cannes Film Festival.

The second is the Regenerative model, one that is gathering international momentum – and which has been most explicitly set out in Kate Raworth's book (*Doughnut Economics*). Raworth's version of the Regenerative model examines the effect of transgressing the two ecological boundaries (planetary and social) essential to maintaining the biosphere. Raworth demonstrates how, through our persistent abuse of the ecosphere, we have severely damaged the social sphere (that is, climate warming has social consequences as much as ecological ones). In this study, I hope to show how, by using this model, we can usefully plot and investigate narratives according to criteria that are no longer genre/star/auteur-led, nor indeed embedded in anthropocentric theoretical models. This Regenerative model will be applied with a view to serving a threefold purpose. First, to demonstrate how narrative cinema colludes with ideological practices of capitalism at work and their global impact (see Chapter 2). Second, to offer fresh readings of certain films within mainstream cinema that have been rather too summarily dismissed as failing to deliver an ethically appropriate narrative (see Chapter 3). Third, to examine in greater detail the narrative construction of two films that give evidence of ecologically sound production practices (see Chapter 4). In all three instances, the hope is that this approach will inspire film practitioners, film theorists and analysts, and film students alike to consider how they might integrate this ecologically based Doughnut model into their work as part of their process.

So let us begin!

1 Film and ecology, Doughnut Economics, and film theory

Cinema is an industry and as such it is an ecosystem. Furthermore, it is also complicit in ecological issues much like other industries. Indeed, a number of studies have examined the cinematic footprint of filmmaking and, too, film and its relation to ecology. And these two interconnected issues – film as an ecosystem and film as an industry – will be discussed in the first two sections of this chapter. As we come to consider film as an industry and how current ecocriticism has raised some very significant issues in this domain, what does become clear, however, is that there has not, as yet, emerged a theoretical model, *per se*, that could help us, on the one hand, give critical readings that are not bound by standard approaches and, on the other, offer the possibilities in film practice to structure film narratives according to different models other than the dominant ones in use today, But, as I hope to show, this is where Raworth's Doughnut Theory comes into play, and I'll begin to present her model in Section 3 of this chapter.

1.1 Film as an ecosystem: film technology then and now – from celluloid to digital

Originally, cinema came down to the revolution of wheels and reels at the right speed in order to create the illusion of real movement in real time. At first, this was a hand-cranked affair: images were made by capturing light onto film, light sources were primarily natural (until studios were electrified), and screened-images were produced by light shining through the processed film. At the source of film technology, then, it is all about light, both a natural resource and a man-made one. With film in this mode, time is linear and the process is analogue. This cinema of the Celluloid era is a Modernist haptic industrial affair.

Nowadays, we have entered a different technology for filmmaking, the digital – which, while still dependent on light, now pixilates visual information. Thus, information, images, and films can be relayed far more readily to us via satellite, the internet, or compressed onto a hard drive or DVD disc. In this new revolutionary context, images come to us at speed in space and time – virtually and instantaneously. Along with its production and distributive potential, digital film is a borderless visual entity, truly transnational, therefore, even as it is simultaneously embedded in a given, specific culture. Thus, in Film Studies, we could speak more helpfully perhaps of film culture, rather than national cinemas – a concept which allows us to explore film as a convergent, interconnected artefact (much as we do in relation to sculpture and painting).[1]

However, we need to remain aware that this digital cinema is a Postmodern, non-haptic, virtual affair. Digital film cannot be seized manually in the way celluloid film can. We can see it, but we cannot grasp it. This inevitably causes a new sort of distance between us as spectator and the object viewed on screen which the plasticity of celluloid film did not produce, because images of the «real» had been recorded onto a *physical* property. But with the digital, this lack of the physical (proof, if you will) and, therefore, the lack of the haptic, means that a kind of emotive annulling can occur, a detachment from a sense of the real. At its most extreme, used in warfare, this micro-digital-technology can create the sensation that what we are witnessing is an illusion, a special effects moment, a spectacle without consequences. Because it is so minuscule (to the point almost of invisibility[2]), this technology allows us to install camera probes in bombs, to send drones mounted with cameras that are armed to drop bombs unmanned. In this context we are truly in the domain of Robotic Digital Practices – what we might term weirdly weightless wars. Thus, for example, during the first Iraq War (known also as the Persian Gulf War) of 1991 and again in the 2003 Iraq War, cameras were mounted in the nose cones of Guided Bomb Units, and, as they steered towards their targets, they provided the (in this instance television) audience with images right up until the final moment of the actual bombing. This smart-bomb footage had the peculiar effect of placing the audience-viewer as the

---

[1] And the question that emanates from the distinction between these two modalities (celluloid and digital) becomes, therefore, «what is their relation to the REAL and to DURATION». A question I leave hanging, because others such as Deleuze (1989), Rodowick (1997), and Marks (2000) have already lengthily debated this issue.

[2] Also the digital images produced can be far more easily morphed, intruded into than celluloid film.

projectile itself, on its way to its pre-programmed destination. But, because upon impact all visual connection ceased, so too there was nothing to see: 'no bodies, no destruction,' 'just a wall of static.'[3] In short, the smart-bomb offered a bird's-eye view of a bloodless war, the whole effect being more like a video-game than anything real. In a similar vein of detachment, I would evoke the relaying, by helmet-cam, of the assassination of Osama Bin-Laden in 2011 to the White House 'Situation Room.'

The interrelationship between war-technology and the film industry is nothing new and dates back almost to the beginning of cinema at the end of the 19th century. Cinematic practice has benefitted from devices designed for warfare starting with the anamorphic lens of World War One (at the beginning of the last century), moving onto the lightweight cameras of World War Two (in the middle of the 20th century); then, the sub-miniature cameras introduced during the Cold War, followed by light, portable TV cameras first used to record events of the Vietnam War; currently, in the so-called war against terrorism we have arrived at nano-sized digital cameras that transmit imMEDIAte information.[4] But the point is *also* that this technology of recording and, thereby, of cinema itself are ineluctably linked to violence; both to our ability to perpetrate it *and* to record and watch it. We watch violence, in both narrative and documentary films, exposing ourselves to the horrors perpetrated on others (at times, to the extent of enjoying being horrified, for example, by slaughter films of any description).

This violence is also linked, as an effect, to our drive to colonise space (first this planet's, then outer-space). The Cold War is exemplary of this, for, in our drive to conquer space, during the 1960s and

---

3 See Roger Stahl's excellent talk 'Through the Crosshairs,' reproduced as a PDF Transcript by the Media Education Foundation. www.mediaed.org/transcripts/Through-the-Crosshairs-Transcript.pdf (accessed 15 July 2019). In it he demonstrates the collusion between camera technology, film production, and the drone war offensive on terror which works to produce what he calls a 'weaponised gaze' of an assenting public.

4 As we know, the use of satellite imagery and drone-mounted cameras is widespread in numerous domains of 'observation' and/or 'surveillance.' A recent example of this technology that could give us heart, however, in that it reveals acts of violation of international standards, is the exposure of Israeli spraying of herbicide near Gaza – thereby depriving Palestinian farmers of crops. A study, using video footage and satellite imagery, 'tracked the drift of herbicides into the Gazan side' of the Gaza strip and 'concluded it was killing agricultural crops and causing unpredictable and uncontrollable damage.' (Miriam Berger, 'Israeli spraying of herbicide near Gaza harming Palestinian crops, report says,' *The Guardian*, 20 July 2019, 35.)

1970s, East and Western blocs sent unmanned, then manned satellites armed with cameras, first, orbiting the planet, then shooting away to the moon, with film being *the* agent of a new colonisation, insofar as it recorded the images of «one small step for (a) man, one giant leap for mankind».

Indeed, such is the revolution of our camera technology (man's material product) that no space is without observation. In effect, we have made all space visible, even though we have yet to recognise fully what it is that we see: namely, that we have rendered all space (be it land, underground, oceans, rivers, lakes, air, space, outer-space – our bodies even) a place, not just of exploration, *but* of exploitation, aggression, conflict, and war, so that nowhere is free from our sight or grasping drive.[5]

This word «grasping» is key in that it directly links to our sense of an inalienable right to «own» the planet (and indeed planets beyond this one). Furthermore, this conquering of space is tightly linked to ideological warfare, namely, the post-war race to prove the superiority of Capitalism over Soviet Communism (and vice versa). But its consequences are deeply linked to ecological matters because at the core of this race, in terms of energy, lies the nuclear. And, in terms of proving the might of one ideology over the other, visual technology was crucial in this struggle. Cameras were used in the former USSR's first satellite orbit of the earth in 1957, providing images as evidence of Russian superiority over the USA. In the mid-1950s, Wernher Von Braun (a leading Nazi nuclear rocket scientist recruited by the Americans) encouraged the then President Eisenhower to pursue nuclear development and to endorse human moon exploration. At this stage of the race, the Russians had the more powerful rockets. In 1959, the Soviet Luna 3 rocket took the first photography of the never-before-seen dark

---

5 It is noteworthy that satellites, since the end of the Cold War in the late 20th century, have increasingly become conduits for military espionage; that, since man's invasion of space, there is considerably greater débris which can cause collisions to occur; that, as acts of censorship by specific countries wishing to withhold information, interceptive rockets are sent up to bring down their own satellites; that, in the same vein, telecommunication satellites can be blocked by countries wishing to censor access to information by its people. Space is now so militarised we can speak of diplomatic tensions in outer-space (the zone beyond the air space/ territorial space surrounding the earth), which technically and legally speaking belongs to no one country! Currently, satellite platforms are for the most part 50/50 military and civil (that is, scientific) in function. The USA leads with some 150 satellites versus the other main countries to have satellites in space, namely, Russia, China, and France with some 15–30 satellites apiece.

side of the moon. Thanks to Von Braun, however, Soviet superiority was soon surpassed by the USA whose own nuclear weapon technology led them to design smaller lighter warheads. And they successfully launched their first satellite, Explorer 1, in 1958. This was later followed by the Apollo Program, and Von Braun's dream to land a man on the moon became reality, in 1969, with Apollo 11.

The effect of this space race was two fold. First, the concept that another planet was «conquerable» was established as fact. This also meant that the pioneering Americans could exploit whatever minerals they found on this new planet to their advantage. Second, the space race very quickly transformed into a star-wars race. Because the Russians had successfully orbited the earth with their rocket-launched satellite (in 1957), this meant that a surprise nuclear attack could easily be launched by the Soviets against the USA (the Bay of Pigs incident, 1961, was a timely reminder of this potential). And so the nuclear arms race became as much a part of the ideological battle as the space race.

On this question of ideological space-race, and to conclude this section, I'd like to briefly pause on a film which stretches these issues into other significant and concomitant domains which, at the same time as these daring exploits were being undertaken in the 1960s, were for their part couched in silence. The film, based on true events and people, is *Hidden Figures* (Theodore Melfi, 2016), and the key word is 'hidden,' or silenced, bleached out. For, the narrative places two race issues at its core: the space race and the question of race and race relations, or, better put, systemic racism. In terms of the space race, on the surface the film's narrative appears to overlook the implications of such an undertaking both in terms of the ecological impact and the uses of the scientific knowledge deployed for war-technology (nuclear arms). To all appearances, the narrative presents us with a POV that endorses the Anthropocene: for there is no *apparent* diegetic questioning as to whether there are any ecological issues at hand, nor whether there are any war implications, nor indeed weaponry developments attached to this technological endeavour (which of course there were). Instead, we repeatedly hear the message coming down from on high, from the White House, that the fight for control of outer-space *has* to happen, America must win. And it will have its heroes, of course, in particular John Glenn. What is certain also is that, during the era in which this film is set and which the narrative reflects, America was hugely engaged in its nuclear programme and, therefore, was not mentally in a place where it would consider, in ecological terms, the threat the nuclear represented to the planet, which was (and still is) total annihilation – even though America was well aware, after Hiroshima and Nagasaki, what the effects of a nuclear fall-out were. But as the next chapter will demonstrate, whilst the nuclear may be the summa of all destructive energies, it is not alone in its damaging of the earth-world.

Back to *Hidden Figures*. If the nuclear threat is «bleached out», silenced, even so, underneath this seeming normalisation of the scientific process of getting a man onto the moon, what does get exposed in this film is a visualisation of the destructive nature of the Anthropocene imperative at work: first, in the form of a technology that seeks, no matter the cost, to colonise other planets as if that were perfectly normative (a continuum of growth economics, and a belief that progress is all[6]), and, second, a visualisation of White masculinity's ruthless, individualistic competitive drive. As just one example, one of the White, male scientists deliberately tries to block the brilliant work of one of the African-American women mathematicians, the so-called Black-computers, who manages to resolve a particular problem that ensures a safe re-entry of the astronauts from space. Questions of race in this context become also questions of gender, social inequality, political voice, income, and education. For, brilliant mathematicians though these African-American (Black-computer) women may be, nonetheless, White privilege determines they shall be shunted away in a back room, unseen, whilst the White men in the Central Command are in «full control». However, the three central African-American women characters' greater knowingness and mathematical intelligence pierce the walls of segregation. Not only do they cross gender and racial lines, in so doing they *expose* the systemic hypocrisy of racism and sexism, always through example, often through silence – indeed, meeting silencing with silence proves to be a powerful tool. As to the space-race question, its colonising paradox notwithstanding, the central point this film makes is that, much as *man* may wish to conquer space – after all, Homo Americanus desperately wants to beat Homo Sovieticus to the moon – he cannot do so, however, without Feminae-Africana-Americana, all three of them:

6 Still this pursuit of colonising planets goes on. Space probes have journeyed to all the planets of the solar system. Colonising Mars is the next target. And currently there are plans, 'from the US, Russia and China, to return to the moon and build a "base" there' first for robots to 'man' and then, in all probability for humans too (Martin Rees, 'The moon was once a frontier. But new worlds now beckon,' *The Guardian*, 20 July 2019, Journal section, 1). China has already explored the possibility of greenhouse technology on the moon to grow crops (www.wired.co.uk/article/china-grows-first-plants-on-the-moon [accessed 27 November 2019]). See also the article on zero-gravity 3D printing, which it is believed will allow for the building of cities in space ('Houston, we have a printer' *The Guardian Weekend*, 21 April 2018, 25–30, no author cited).

Katharine Johnson has the maths, Mary Jackson has the engineering know-how, Dorothy Vaughn has the brilliance to make IBM work. Indeed, it could be argued that by placing this Black-race story within the White-space-race story, that is, within the White all-male world of scientific privilege, this film exposes an «unknown» (to whom?) piece of history. But it also exposes the symbiotic interrelatedness of a colonial mentality that can unquestioningly continue its march forward to own more space and, in the same breath, walk asunder over sections of humanity it considers of lesser value than itself.

1.2 Cinema as an industry: eco lit and film

This will be a brief introduction to some of the issues already raised in relation to film and the environment (the Annotated Bibliography at the end of this book provides further references). An obvious starting point (now that it has been brought to our attention) is the impact of the film industry itself on the environment. To date, the economies of the industry have been integrated into film studies, but primarily by looking at budgets and revenues. Only recently have we begun to think about the industry's carbon footprint. Here, just in terms of the film shoot, are a few matters to consider: air-miles accumulated, damage to local environments to create the landscape desired by the production company, the detritus left behind in location shooting (to say nothing of the impact of generators used to keep lighting, cooling, and so forth going, nor indeed the huge number of plastic containers used for feeding crew and cast). It is the case that production companies are now off-setting their carbon footprint. For example, since the early 2000s, Hollywood has been «neutralizing» its carbon emissions by paying off their «dirty» practice through donations to forestry departments. However, the damage has been already been done.[7] A preferable approach would surely be to reduce the carbon footprint in the first place, and to exercise greater restraint when on location shooting. In terms of off-setting, the question becomes: how is it possible to imagine that the planet's well-being can be tied into a capitalist model of credit? Clearly, the film industry still has a long way to go in its endeavours to become green(er). Globally, the audio-visual sector's (film, TV, Radio) annual

---

7 Danny Boyle's *The Beach* (2000) is an early example of this monetary off-setting practice (one is tempted to call it guilt-credit). *The Day After Tomorrow* (2004), *Syriana* (2005), *Evan Almighty* (2007), and *2012* (2018) are just a few films we can mention to have used this method of carbon offsets rather reducing their own emissions.

carbon footprint is 1.1m tonnes of CO_2, a quarter of which is directly attributable to the actual filming. Moreover, Hollywood is still the second most polluting industry in California.[8] The answer to some of these issues is to film more locally, use local resources, and employ personnel who are willing to travel on a more ecological-sensitive basis (as some technicians and actors, including a handful of film stars already do). In Europe a new initiative has been launched, ECOPROD, whose function is to sensitise the audio-visual sector to a 'better consideration of the environment' and to do so by providing a 'professional resource center to reduce the environmental footprint (...) through free tools such as fact sheets, stories, guides, directories, follow-up productions.'[9]

Much has been made of the positive impact on film production of the move to digital practices. Indeed, there has been a huge diminution in costs: from $2,000 per print of film to $50. We can also consider the low costs of cameras and editing practices and the implications of this for poorer nations who now find themselves in a position – at least in technical terms – to make more films. Finally, quality is maintained during a digital film's screening lifetime. However, it too has its carbon footprint. First, in terms of the actual shooting of the film, the problem remains much the same as ever (see above). Second, there are environmental impacts of the actual technology which make the digital non-eco-friendly, even if it is not to the degree of its earlier celluloid prototype. With celluloid production and processing, a filmmaking factory (such as Kodak) used huge quantities of petrol and chemicals to make and process their product, to say nothing of the massive amounts of freshwater for cleansing purposes which then got sent back into rivers (thus polluting them). For its part, though, digital cinema encounters more than mere run-on costs such as storage,[10] cooling systems, and updating existing digital films to new software. It also produces a sizeable carbon footprint in terms of the accumulation and disposal of electronic hardware. Existing hardware rapidly becomes obsolete; that waste is then shipped out to poorer nations (cheap labour) to have its useful materials recycled; and the rest is thrown away, which, in turn, adds to the environmental problem of material seepage into the earth and rivers.[11] For more information on

---

8 Listen to France-Inter's news programme *Le journal du 7/9* www.franceinter.fr/emissions/l-edito-carre/l-edito-carre-03-octobre-2019 (accessed 03 October 2019).
9 See Ecoprod's manifesto on www.ecoprod.com/en/ (accessed 03 October 2019).
10 Archiving digital masters is more expensive than storing celluloid film masters.
11 On these specific issues, see Maxwell and Miller's excellent chapter 'Film and Environment' in Hjort (2012: 271–89).

these issues, I would suggest looking at two very useful studies: Nadia Bozak's (2011) *The Cinematic Footprint: Lights, Camera, Natural Resources* and Mette Hjort's (Ed., 2012) *Film and Risk*.

Cinema as an industry is 'resource-derived and energy-driven' (Bozak, 2). As such, it is an ecological practice. It is also an ecosystem. That is to say that all films are produced in relation to the material world; thus, each and every film creates an environment which we, as audience, consume with varying degrees of awareness as to the historical, political, social, and ideological effects implicit in that viewing experience. A considerable number of studies have focussed on this concept of film as an environmental relationship. In *Screening Nature: Cinema beyond the Human*, Pack and Narraway ask that film studies move away from films that 'corroborate human exceptionalism' and look, rather, to explore the 'convoluted relations between humans (…) and non-humans,' in short to examine more carefully the interdependence between 'the natural world and humans within it' (2013: 7). Narrative films and nature documentaries *can* achieve these goals. But, as the editors of *Ecocinema Theory and Practice* point out, many documentaries about the planet (particularly those made by Hollywood, or for television) still adopt an anthropocentric approach, and most fiction films tend to see landscape and environment as a back-drop rather than an integral and active agent within the narrative (Rust et al., 2013: 7–8).

The call in these critical studies is for a more ecocentric approach to cinema, to go «beyond the human» in our analyses; to ask the question, how much does cinema suppress the non-human elements? That is to say, there is a need to recognise that the planet is constituted of both the human and the non-human (animals, landscape, nature, environment); both must be acknowledged in film theory. Ecocriticism in this context is strongly linked to posthumanist theory of which there are (at least) seven definitions, two of which interest us here. The first, as you can infer from the above, is one in which posthumanism relegates humanity back to one of the myriad of natural species. Logically, it follows that anthropocentric dominance is an aberration, and a practice that must be rejected. This is the definition adhered to by ecocriticism. The second definition of significance, however, is virtually the complete opposite of the first. Posthumanism in this context is perceived as an intensification of humanism – that is to say that science and technology have yet to «perfect» the human. The belief is that we can go further in our species' development, aided by modern scientific tools (such as A.I., brain implants, etc.). In this modality, posthumanism is closely linked to the concept of transhumanism. Presently, it is this latter definition (how we can progress, develop, and transcend our potential) that seems to hold greater ontological sway – but not just in films, it is also very present within the real world (a point we'll be returning to when we consider the impact of capitalism both on our world view and its representation in film in Chapter 2).

The key to looking at cinema through a new optic, one that takes on board issues raised by ecocriticism, is to recall, first, that cinema is not *apart from* but *part of* the ecological system and therefore has its own effects on the planet. Equally important, for any theoretical model, is the breadth of the problem in that, over the last 50 years or so, the planet has been hugely impacted upon by the effects of globalisation. As the power and dominance of the West has diminished and the accelerated modernisation in new emerging economies (such as China and India) has increased, to say nothing of the effects of post-colonisation, global warming, and migration, so too we are spiralling faster and faster towards our own, if not the planet's extinction, as resources diminish almost daily. But, as the editors of *Ecocinema Theory and Practice* point out, adopting a moralistic tone within ecocriticism will not work in the main; film can challenge our 'ecological awareness,' but it can just as easily be 'completely ineffective' (5–6). Nor will a narrow definition of ecocinema serve our purposes. Rather, a holistic ecotheory – one that takes on board the material, social and perceptual earth-world – is what is needed (6).

Bearing the above problematics in mind, and by way of a concluding section to this chapter, I'd like to give you a foretaste of a model which, I believe, answers many of the problematics raised above. What follows is a brief, explanatory outline of Kate Raworth's model which she calls 'Doughnut Economics.' We shall be returning to it in more detail throughout this book. But I thought it would be helpful to put it in place here so that, as you read the intervening chapters, you can have this model in mind as an inspiration and a hope that all is not yet lost!

1.3 Doughnut Theory

The original meaning of economics, Raworth reminds us, is household management (*oikos nomos*), which in today's world means our duty of care to our planetary household (2017: 4). Why not, she then asks (43), focus on advancing the richness of human life rather than the richness of the economy (which after all is not a sentient being). To this effect, she draws a different economic model (see Figure 1.1) to the ubiquitous growth one, bringing together the two spheres of true economics which are:

1. The planet economics/planet ecology
2. The social economics/social ecology

14 *Film and ecology*

Figure 1.1 Raworth's basic Doughnut diagram.

The first sphere encircles the second to produce what Raworth terms the Biosphere: as we can see, in diagram form below, sphere one – planet economics – encircles sphere two, the social economics, and the space between the two spheres creates the green Biosphere which is Raworth's term for the ecology-sphere, the space of planetary and human safety (11).

Raworth goes on to explain that to transgress these two economic ceilings/boundaries leads, in the first instance, to critical planetary degradation by overshooting the ecological ceiling (seen in red on the diagram), and, in the second, to a socially unequal and unjust space by falling short of minimum social standards (also in red in the diagram). For its part, the Biosphere revolves and recycles, regenerating planetary and social life, thus providing a safe and just space for humanity (44). In a more detailed diagram, Raworth expands on the present situation in which we find ourselves and in which evidence clearly points to how extensively the Anthropocene has overshot and fallen short of minimum standards. As follows (Figure 1.2):

Film and ecology 15

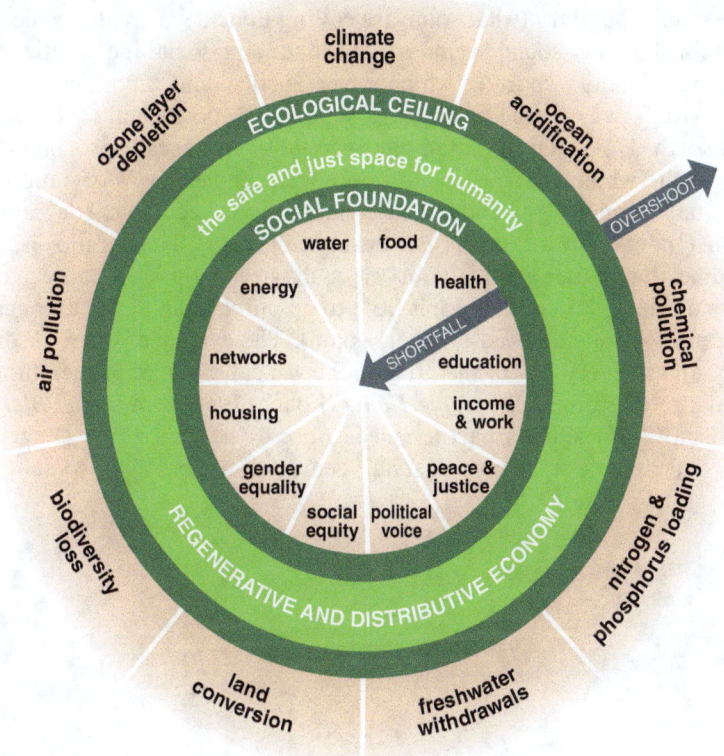

Figure 1.2 Raworth's detailed Doughnut diagram (2017: 51).

To explain: the environmental ceiling consists of nine planetary boundaries (all written on the outer perimeter), beyond which lies unacceptable environmental degradation and potential tipping points in the Earth systems. The 12 dimensions of the social foundation (written on the inner perimeter) are derived from internationally agreed minimum social standards, as identified by the world's governments in 2015. Between the social and planetary boundaries lies the Biosphere, an environmentally safe and socially just space in which humanity can thrive (Raworth, 44). But as you can see from all the red-orange heating, we have overshot and fallen short to a concerning degree.

Thanks to this detailed diagram we can also see how the one sphere of economics is impacted upon by the other (and vice versa). In brief: the abuse of one foundation entails the abuse within the other. As Ai Weiwei's documentary (*Human Flow*) makes clear, migration is the new condition of globalisation *and* of insidious ecological mismanagement. Indeed, the cumulative effect of poor planetary management on the outer boundary (what some are calling Ecocide[12]) creates a shortfall for the inner social boundary, thus inducing social inequalities of all orders, and a space in which peace, justice, and a political voice are withdrawn. As the economists Wilkinson and Pickett (quoted in Raworth, 171–2) have pointed out 'inequality damages the social fabric of the whole *of* society' and, moreover, 'higher levels of national inequality tend to go hand in hand with increased ecological degradation' (India being a prime *visible* example of this[13]). The paradox is, of course, that in the end, inequality slows down economics because markets diminish (after all, there is little to no purchasing power amongst the mass of unemployed or poorly paid). The evidence is here in the UK and in every city of the UK, and as we can observe in such films that stretch from *Cathy Come Home* (Loach, 1966) to *I Daniel Blake* (Loach, 2016) and its partner piece *Sorry We Missed You* (Loach, 2019), for the poor, over the last half century it has only got worse.

---

12 See the work of Ecocidal Defence Integrity, a movement founded by ecological lawyer, the late Polly Higgins.
13 It is worth mentioning that India's poor are hitting back with documentary films of their grievances (via Video Volunteers).

2 Film and the Anthropocene
Dirty Capitalism – *Mildred Pierce* (1945), *Tulsa* (1949), and *Giant* (1956)

2.1 Anthropocene visualisation – Baudrillard's nightmare – A.I.'s death wish, robotisation, and anthrocide or beautiful destruction

In relation to the visualisation of the planetary degradation in which we find ourselves, art historian and cultural critic T.J. Demos (in his book *Against the Anthropocene*) poses the central question: as opposed to the current tendency to show the surface reality of disasters and not the subsurface damage (be it planetary or social), how can we convert into image and narrative the disasters that are on the whole (until they explode in our faces), slow moving, long in the making and often invisible (2017: 13)? Disasters that, whilst created by agro-business, petro-business, geo-engineering, drug and chemical industries, bankers, and investors, remain «anonymous» and «star» nobody; disasters that are attritional – the effects of what Demos calls slow violence – but which are of little interest to the sensation-driven technologies of our media-image world (13).

Furthermore, Demos asks (37) how to combat images that work towards reassuring us of the controllability of disasters (such as oil spills). Such visualisation of reassurance with regard to real disasters finds very similar echoes within narrative film and its three-arc narrative, which runs as follows: a disaster has occurred, we are caught up in the maelstrom, but human ingeniousness will put it right – with the ingenious element often embodied in the form of one heroic white male. In these films, we go to the brink and pull back/escape just in time. Equally concerning, the disaster is rarely sourced to Anthropocene activity. As a simple example of high scoring in denialism, consider this: out of a 1,000 Disaster Movies listed, the highest percentage are labelled «Natural Disaster Movies» at 27%, with «Man Made Disaster Movies» at a mere 9%. So, de-responsibilisation reigns. Furthermore, for every film-disaster narrative, whilst for the most part the disaster takes everyone by surprise, there is nearly always a quick-fire solution to it. Our difficulty, it would seem, is our inability to accept that planet degradation is a matter of slow violence – almost too slow for us to recognise until it is too late, as in the case of the Arctic ice-cap melting, for example. Amongst the few exceptions to this narrative pattern we can count Kramer's post-apocalyptic film, *On the Beach*, 1959, where indeed it is too late; and perhaps, in a different vein, Kubrick's exposure of the tyranny of not understanding our technology in his satire on A.I. and modern technology, *2001: A Space Odyssey*, of 1968, one year, incidentally, before the moon landing, lest we forget.

Finally, in this context of Anthropocene Visualisation, *what* about the invisibilities which technology can create by manipulating digital information and making the «problem» disappear, go under the surface as if it had never «really» occurred (Demos, 36)? A major example of this is the geo-technological disaster in 2010 of the BP Deepwater Horizon oil spill in the Gulf of Mexico – the worst marine oil spill in history; and still not resolved, the invisible continues to take its toll – not least of which the effects of the toxic chemicals used to try and clean up the ocean; indeed, instead of so-called «fixing things», it's a question of doing more of the same and, as is often the case, making it worse (Demos, 51). A 2016 faction film *Deepwater Horizon* (Peter Berg) covers much of the events of this disaster, including the cost to human life, and the lack of accountability meted out by the courts to those responsible. The film remains, however, a story told on the *human* rather than the *planetary* level, as are most disaster movies.[1] The film focusses almost exclusively on the bravery of the men on the oil rig, led by a very determined hero who fails in his efforts to counter the stupidity of the on-board management bosses whose ignorance of drilling dangers, coupled with their blind subservience to their oil-company masters, culminates in the disastrous blow-out of the rig. This binary tension between hero and corporate idiots, whilst significant, falls short of the major issues at hand. First, it fails to spell out the magnitude of the disaster and its cost to the environment. Second, it does not question the practice of oil-drilling per se, the very industry of destruction.[2]

---

1 It may interest you to know that a mere five years later, in 2015, BP, in a bid to take advantage of climate change/warming, filed to extract oil from the Arctic Ocean!
2 A number of European company pension schemes are presently withdrawing from investments in companies that violate the Paris climate agreement. (See: Patrick Collinson 'Why is it so absurdly difficult to stop your pension money worsening the climate crisis?' *The Guardian*, 10 August 2019, 42.)

Deepwater Horizon serves a different purpose to the one proposed by Demos. It is generically caught in a trap of its own making. The trope of a disaster movie is both to terrify and to lay bare human failings (in this case hubris) as well as its opposite (courage) – all at lightning speed. As a visual spectacle, therefore, there can be no mise-en-scène of the slow-violence Demos speaks of. Disaster movies don't have time to stop and question. Technological pyrotechnics is their game. It is the case, as Baudrillard has pointed out, that we appear to be in awe of our own sublime technology, to the point – at least in films – that we almost take pleasure in our own destruction (*The System of Objects*, 1996: 122). In this context, disaster, science-fiction, or futuristic films act as a projection of our aesthetic pleasure in our own (near) annihilation. This time, as opposed to the slow violence of real disasters (discussed above), the aesthetics of violence are fast moving, choreographic even. In the main, science-fiction and futurist films project Robotic, Android, or post-nuclear dystopias as the enemy that can be defeated, or a «somewhere» where human life will somehow still exist (even if by a hair's breadth, or a single birth!). Even such undoubtedly well-intentioned films as *The Day After Tomorrow* (2004), which exposes the dire effects of climate warming, run into similar narrative problems through simplifying both the events caused by the Arctic cap melting and the end message of the film: the climate denialist President is converted, young survivors will carry forward new generations. It is worth remarking that, for the most part, these narratives overlook the fact that these very conditions are of our own making, that the responsibility is a collective as well as an individual one. The tendency is to reduce the plot to a duopoly between good and bad Anthropocene (see, for example, *Mad Max Fury Road*, 2015). But rarely are the exploiters brought to task. This is made clear in the nuclear-disaster movie *The China Syndrome* (1979). Courageous as the central characters are in preventing the nuclear power-plant disaster, there is every evidence that there will be a cover-up, that the negligent senior executives will not be brought to task and the company will continue to cut corners.

And yet, as a handful of science-fiction films remind us (for example, the 1968 *Planet of the Apes*), nature *is* the source of life, not man/Anthros. Even more remarkable, given its timing of 1951, is Robert Wise's *The Day the Earth Stood Still*. Consider its narrative: against the background of the Cold War at its first peak (the Korean War), the beginnings of the Nuclear Arms Race (between the USA and the former USSR), the arrest of US spies Fuchs, Hall, and the Rosenbergs – against all of this, this film delivers an extraordinary message. A man,

Klaatu, lands from outer-space in an endeavour to stop the escalation of war-technology on earth; his own planet is astonished at the madness of our world in which nations do not get along with each other and in which man has harnessed nuclear energy, but only so he can use its killing power. Klaatu's mission fails, but as he takes his leave in despair, he delivers this message: 'Security for all must be a planet without aggression, your ancestors knew this' (it seems we have forgotten).

2.2 Dirty Capitalism

As T.J. Demos (97) explains, the choice is starkly clear. On the one hand, we can invest in the formation of re-localised sustainable cultures based on renewable energy systems; we can practise *de*-growth and redistributive economics, thus ensuring climate justice and post-capitalist democratic practice. In so doing we acknowledge the dynamic instability of the planet – we accept that life (as Stuart Hall puts it[3]) is a state of permanent revolution, in the sense of a rotating regenerating revolution. *Or*, on the other hand, we can continue down the Anthropocene-Capitalocene path of extreme geo-engineering in an age of climate change ruled by centralist, increasingly authoritative governments and their repressive militarised police: witness Trump and his new off-shore oil and gas leasing programme, the land grabs off Aborigines in Australia documented in John Pilger's *Utopia*, 2013, the 2017 Myanmar military genocidal clearance operations against the Rohingya Muslims in order to take over the petrol and gas-rich territories in Rakhine State, or again, President Bolsonaro's drive to remove, by any means necessary, Brazil's indigenous tribes from their territories, a prime target for exploitation being the Javari Valley rainforests. We need only document these few examples of autocratic disregard and map them upon Raworth's doughnut diagram to see how clearly this is a path that leads to increased forms of socio-economic and political inequality. These strategies break every single one of the nine planetary boundaries and bring in their wake an abuse of the 12 dimensions of the social foundation (see Figure 1.2). In Brazil, with the continued, often illegal, assault from loggers, ranchers, hunters, oil firms, and gold-diggers on their territories, the health of existing indigenous tribes, living in voluntary isolation and therefore lacking

---

3 Stuart Hall: 'We are all living in a state of permanent revolution.' Cited in John Akomfrah's documentary on Hall, *Unfinished Conversation* (2012).

immunity, is completely vulnerable to the simplest of illnesses[4]; for over a century Australian Aborigines have been stripped of their human rights and forced into living against their traditions with all sorts of social and health consequences; the surviving Rohingya, having first sought refuge in Bangladesh, are now destined to be moved to Bhasan Char, an island at risk of floods and cyclones – all examples of displacement for capital gain and indifference to human well-being.[5]

So let us pause here on what I term Dirty Capitalism, observe it in relation to the above diagram, and see where it has presently led us to.

To begin: on 6 May 2019 the UN Intergovernmental Science-Policy Platform on Biodiversity and Ecosystems Services (IPBES) published its report in which it detailed explicit evidence that the world is facing its sixth wave of extinction, but, unlike the past five, this one is driven by humans. According to this same UN report, the chief causes for nature's collapse are the conversion of forest, wetlands, and other wild landscapes into ploughed fields, dam reservoirs, and concrete cities. To date, three-quarters of the world's land surface have been severely altered. The Chair of IPBES, Professor Robert Watson, is very clear that the 'ones responsible for the drivers of biodiversity loss' are 'those in charge of agriculture, transport and energy.'[6] If you refer back up to Raworth's Doughnut diagram (Figure 1.2, p. 15) you can see how these industrial systems, along with the petrol and chemical ones, have put the ecological ceiling in great danger.

4 Re: the illegal logging practices endorsed by President Bolsanaro see, Tom Phillips, 'Bolsanaro has given his blessing to brutal assault on Amazon, sacked scientist warns' (*The Guardian*, 10 August 2019, 30). And on the indigenous tribes at risk see, Tom Phillips, '"Bolsanaro wants to destroy the lot of us", Indigenous tribes of the Amazon prepare resistance to invaders' (*The Guardian*, 27 July 2019, 33).
5 It may interest you to know that the Rakhine State is not just a strategically important territory for the Myanmar government, but also to its neighbours China and India. The discovery of massive oil and gas reserves along the Rakhine coast in 2004 and its subsequent exploitation since 2014 is a big contributing factor to the removal of the Rohingya. A move supported by both the Chinese and Indian governments. Of the 400-m cubic feet of gas produced every day, 379m are exported to China. The 771-km pipeline that transports oil from the Myanmar port of Kyaukphyu in Rakhine to Kunming in China brought about the removal of the coastal Rohingya community in 2012. See Zoglul Kamal's article 'The untapped wealth of Rakhine and the persecution of the Rohingya,' www.dhakatribune.com/world/south-asia/2017/09/26/untapped-wealth-rakhine-persecution-rohingya/ (accessed 28 July 2019).
6 Watson quoted in *The Guardian* (4 May 2019, 16). Read the full article by Jonathan Watts, 'Earth's life support system close to crisis point, UN report to warn.'

But let us be more specific and take a look at two contrasting stories within the domain of agriculture – the one clearly implicated in negative effects, the other, one would have believed, the direct opposite. As we shall see, both have had negative impacts on the environment and the social sphere, pointing to the fact that ecological sense must drive our biospheric practice well before economic gain. First up is Monsantano, the chemical mega-corporation and producer of glyphosate (now owned by the German chemical giant Bayer). Recently, Monsantano has been taken to court in Oakland and San Francisco and been successfully prosecuted on three occasions for the cancerous effects of Round-Up Ready herbicide (whose active ingredient is glyphosate). These three Californian trials are the first of an estimated 13,000 plaintiffs with pending lawsuits all over the USA. Of course Monsantano/Bayer will launch a series of appeals that will delay any pay-out (so doubtless many of the plaintiffs will die before any compensation is made).[7] Two things emerge from this story. First, that big capitalist corporations can eternally delay any pay-out due to victims – so their responsibility in this form of Ecocide is never or rarely acknowledged (as was well documented in the Bhopal Disaster of 1984 and Union Carbide's protracted defence, with any survivors of the victims of the disaster finally receiving minor compensation in 2004, by which time the story had long evaporated from global consciousness). Second, since it is evident that human health is severely affected by the inhalation of chemical leaks, clearly the question arises as to what happens to other levels of biodiversity.

The other story is differently concerning, because the intention was to enhance biodiversity. Some 20 years ago, Chile, in an effort to increase pollination, imported European bumblebees known for their fantastic pollinating skills. It was not until recently, however, that it was discovered that these industrially bred bumblebees transmit pathogens killing the native bees. But the story is not limited to Chile. Each year more than 2-m bumblebee colonies are exported from factories in Europe to greenhouses in over 60 countries. Once it escapes from the greenhouse, this highly adaptable bumblebee wreaks havoc on indigenous ecosystems. To counter this first disastrous effect – which we might well have considered (initially) to be a wise ecological solution – a Japanese pesticide producer has proposed spraying growth-regulator insecticides on foraging bees who then take the

---

[7] See article by Paul Elias 'Couple win billions in damages over weed killer cancer claims' (*The Scotsman*, 15 May 2019, 23).

poison back into their nests.[8] So a double-whammy effect is the outcome of good intentions: eliminating indigenous bees, then toxically eliminating non-indigenous bees, with doubtless the consequential damage that toxic spraying will bring about on the environment. Ecology requires careful research practice, not quick-fire or superficially self-evident solutions (a practice which, ultimately, is more identifiable with the growth-economy model of current capitalism).

The outcomes of Dirty Capitalism are everywhere to be seen – and they have a long historic tail (the slow-violence Demos speaks of). In India, for example, the building of the Sardar Sarovar Dam, begun in 1961, has brought about the destruction of a whole ecosystem: communities, farms, soil, water, forestry, fish, and wildlife. Over a 50-year period, the protests of the indigenous peoples were repeatedly brushed aside, and despite numerous court cases brought to try and halt the construction, the dam was finally inaugurated in 2017 amidst great pomp. The cost to the ecosystem was huge, but the profits to big business and government were about to become even bigger. The rural, indigenous people were obliged, against their will, to choose: stay and drown or become refugees – a choice documented in Franny Armstrong's 2002 documentary *Drowned Out*. Forced to move and live elsewhere with a minimum of compensation, they barely survived. Whereas before they had a sustainable life, they now became the migrant poor. As Arundhati Roy put it so eloquently in her 1999 blog *The Greater Common Good*:

> Big Dams are to a Nation's 'Development' what Nuclear Bombs are to its Military Arsenal. They're both weapons of mass destruction. They're both weapons Governments use to control their own people. Both Twentieth Century emblems that mark a point

---

8 See Alison Benjamin 'Giant bumblebee at risk: Imports of European pollinators devastate Chile's "flying mice"' (*The Guardian*, 4 May 2019, 17). Urban bee-keeping hobbyists, well-intentioned as they may be, are also causing problems; see, Kate Connolly, 'Berlin's boom in urban beekeeping leaves swarms without homes' (*The Guardian*, 10 August 2019, 31). On poisoning as a bad practice see, Naaman Zhou, 'Poisoning suspected as dying birds fall from sky,' where dozens of corella birds fell from the sky in a case of mass poisoning in South Australia. The poisoning is slow in its killing (up to a week); the intended targeting of the little corella bird (guilty of damaging crops and infrastructure, apparently) did not spare the species-protected long-billed corella bird (*The Guardian*, 13 July 2019). Southern Australia seems to be having something of a culling field day, recent reports tell us of 'animals including koalas and kangaroos' being 'culled in parts of southern Australia, where large numbers are damaging the landscape.' (See, Lisa Cox, 'Overabundant kangaroos and koalas face cull,' *The Guardian*, 13 July 2019).

in time when human intelligence has outstripped its own instinct for survival. They're both malignant indications of civilisation turning in on itself. They represent the severing of the link – the *understanding* – between human beings and the planet they live on.[9]

And, some 20 years later, Roy in her collection of essays, *My Seditious Heart* (2019), is once again making the same points, showing how little the Capitalocene is willing to listen, let alone understand when she says: 'capitalism's gratuitous wars and sanctioned greed have jeopardised the life of the planet and filled it with refugees.'[10] She writes about the impact of the so-called growing economy of India since the turn of this century, when India – having joined the BRIC economy – opened up its markets to international finance:

> India's fleet of new billionaires and its consumers were created at an immense cost to its environment and to an even larger underclass. (...) Labour laws were dismantled, trade unions discarded. The state was withdrawing from its responsibilities to provide food, education and health care. Public assets were turned over to private corporations, massive infrastructure and mining projects were pushing hundreds of thousands of rural people off their lands into cities that did not want them. The poor were in freefall.[11]

Roy's closing vital question is: what will become of so many surplus people? David Hare's 2014 play *Behind the Beautiful Forevers*, based on the 2012 novel by Katherine Boo, about people barely surviving on a slum economy, recycling the mounds of rubbish on the outskirts of Mumbai, provides us with a stark visual idea.

Emerging economies such as India and China, in joining the Neo-Liberalist hegemony that came about in the West with the «Fall of Communism» in 1989, have remodelled it to suit their own political-cultural system and ended up with an even more brutal model than that which

---

9 www.narmada.org/gcg/gcg.html (accessed 26 July 2019). Sadly, dams as weapons of destruction are not limited to this single case. Recently, in Brazil (another BRIC economy) there have been two dam disasters, the Mariana Dam in 2015 (killing 19 people) and the Brumadinho dam in 2019 (with 248 deaths) – both as a result of weak regulatory structures. Moreover, metals in the tailings of these collapsed dams will be absorbed into the rivers' soil and affect the whole local ecosystem.
10 Arundhati Roy, '*My Seditious Heart*' (*The Guardian Review Section*, 01 June 2019, 15). Full reference: 'Extract: My Seditious Heart' (13–15).
11 Roy (2019: 13).

presently dominates most Western democratic economies. But all have in common the two basic tenets of neo-liberalism: recognition (who deserves the rights to capital) and distribution (who deserves income).[12] This, if you will, boils down to saying that, in the first instance, there are those whom hegemony determines merit capital (in short, a system of meritocracies or status hierarchies) – institutional models, examples of which in the West are Wall Street, The City, Silicon Valley, Hollywood, Investment Banks (etc.). And, in the second instance, there are those whom hegemony determines merit the allocation of capital, beginning, most obviously, with those who work, but also those in need (in nations where the welfare state exists in some form or another) – in short, a system based in class divisions.

Let us briefly consider the impact of this «Fall of Communism». Almost immediately upon the Soviet collapse, the economist Francis Fukuyama spoke of the triumph of capitalism over communism as the 'end of history.' In his book of the same name, published in 1992, he boldly asserted that capitalist democracy represented the one ideology that would endure (hence his term 'the end of history' – the end point of man's ideological evolution, apparently!). His claim was that liberal democracy and free-market economy had won over other ideologies.[13] One of the pillars of the Cold War had finally crumbled. The major effect was that free-market neo-liberalism could fully take off, and did so, in the form of the 'dismantling of barriers to, and protections from, the free movement of capital' (which, as we now recognise, facilitated mass money-laundering by multi-national oligarchs from Russia, China, India, and the USA, just to name four countries), to bank deregulation and the 'ballooning of predatory debt' (most readily exemplified by the sub-prime mortgages, and easy credit to persons of small means).[14] Even the global financial crisis of 2007–8 failed to temper the march of free-market neo-liberalism. Meantime the newly nominated BRIC nations (Brazil, Russia, India

---

12 See Nancy Fraser 'From progressive neoliberalism to Trump – and beyond' (2017, 1–18) https://americanaffairsjournal.org/2017/11/progressive-neoliberalism-trump-beyond/ (accessed 25 July 2019).
13 In his book, *The End of History and the Last Man*, Fukuyama writes:

> What we may be witnessing is not just the end of the Cold War, or the passing of a particular period of history, but the end of history as such. (...) That is, the end point of mankind's ideological evolution and the universalization of Western liberal democracy as the final form of human government.
>
> https://en.wikipedia.org/wiki/Francis_Fukuyama (accessed 26 July 19).

14 Nancy Fraser (2017, 4).

and China) – so called by Goldman-Sachs in 2002 – were also on the march with their own forms of neo-liberalism. In all their manifestations, however, these neo-liberal economic policies have succeeded in hollowing out the living standards of the lower classes (and in some countries the middle classes) while 'transferring wealth and value upward – chiefly to the one percent, (…) but also to the upper reaches of the professional-managerial classes.'[15]

There is a certain irony in the fact that not only did Fukuyama appear to overlook the fact that China and North Korea, with their own particular brand of Communism, did not collapse, but that both, in their differing manners, have managed to meld their ideologies with a particular type of state capitalism. Especially ironic now, since, in the West, all evidence points to the fact that the second pillar of the Cold War, capitalism, is falling – certainly crumbling. Everywhere we turn, it appears that political rot is assailing capitalist democracies. Pundits speak of the rise of populism; certain rulers speak of presiding over illiberal democracies (as if it were a perfectly natural phenomenon and not a contradiction in terms). We are told, by world leaders of the West, that we are surrounded by fake news. Indeed, to all evidence, we live in a post-fact, post-truth world. If only it could be a de-growth, post-work[16] world instead… I know, currently, it's unimaginable, a heresy even, but will the unthinkable become the new orthodoxy…?

The end of Communism, whilst it might well have sundered the USSR, has not entailed the end of major global conflict which, presently, in many instances, takes place within a virtual world – the world of algorithms, A.I., data-processing, exploitation of Social Media, and so on. Economists are now writing about Postcapitalism, philosophers, and A.I. geeks about Posthumanism. There is a belief in science that biotechnology can improve the human – that there are means whereby we can transcend human limitations. Genetic modification has already proved that we can overcome certain types of cancer, for example, as with the development of Car-T-Cell technology. The

---

15 Nancy Fraser (2017, 4).
16 A post-work world is one in which all labour is valued. See the in-depth article on Post-work by Andy Beckett, 'Post-work: Is the job finished?' (*The Guardian*, 19 January 2018, 9–11). As Beckett says (citing Benjamin Hunnicutt), the work-ethic is a recent construct and 'contrary to conventional wisdom, work ideology is neither natural nor very old' (10). Post-work would entail a more balanced environment between the domestic, the communal and shorter working hours in the public sphere. Remember John Maynard Keynes in the 1930s predicting that by the 'early 21st century, advances in technology would lead to an "age of leisure and abundance", in which people might work 15 hours a week' (Beckett, 10).

development of 3D-bioprinting means that scientists can produce living organs – surely the summa of what Baudrillard called 'simulacra,' based as it is in a combination of human tissue with bio-mimicry. Life can be improved upon and extended.[17] Next stop transhumanism: not just cyborg-like prosthetics, but existing organs such as eyes, replaced by techno-organs, digitised fingers that can now function as a cellphone, a brain implanted so it can be hooked into the internet by the Home Office (see, for example, the transhuman daughter, Bethany, in the BBC 2019 television series *Years and Years* – a fitting warning for what is to come[18]). In fact, the technology for brain control is almost already here. Neuro-meshing, which is used as an anatomic solution for the reconstruction of cranial defects is at trial stage in mice. After injecting the mesh, which is embedded with nano-scale electronic devices, brain activity is then tracked on a computer.[19] Even Fukuyama, in *Our Posthuman Future: Consequences of the Biotechnology Revolution* (2002), has recognised that biotechnology, because it allows humans to control their evolution, may allow us also to alter human nature, thereby putting neo-liberal democracy in danger!

In this digital, non-haptic age, democracy is indeed greatly at risk, but not, unfortunately, the *practices* of neo-liberal capitalism. As neo-liberalism – particularly the regressive, reactionary form – expands into the newly emergent economies, what was already harmful to the greater good of humanity in Western capitalist countries is now being doubly exemplified in the BRIC nations in that their practice impacts both on their ideology *and* back upon the West. As the Chinese, exiled author Ma Jian explains, the current President, Xi Jinping 'oversees a hi-tech totalitarian economic superpower that poses grave threat to democracies.'[20] For all that 'China might have draped itself in a coat

---

17 Recently, stem-cell scientists have produced monkey-human chimeras, which in their view could address the lack of organs for transplantation and organ rejection. Scientists believe organs genetically matched to a particular human could one day be grown inside animals. This research is being conducted in China to avoid legal issues (for virtually all countries ban this type of stem-cell/embryonic research, Japan, and China being the current known exceptions). See Nicola Davis 'Scientists produce monkey embryos with cells of humans' (*The Guardian*, 03 August 2019, 16).
18 *Years and Years*, a six-part TV drama series, by Russell T. Davies, first broadcast on BBC 1, 14 May 2019–18 June 2019.
19 See: www.extremetech.com/extreme/207848-injectable-neuro-mesh-covers-the-brain-can-control-individual-neurons (accessed 01 August 2019).
20 Ma Jian, 'Surviving Tiananmen' (*The Guardian Review Section*, 01 June 2019, 37) (full page reference 36–7).

of prosperity, (...) inside it has become more brutal than ever, and it's this venomous combination of extreme authoritarianism and extreme capitalism which has infected countries around the world.'[21] Yet world leaders 'continue to appease China's autocrats for their own countries gain.'[22] A vicious circle ensues whereby increased executive power coupled with a disregard for the truth has infected so-called democracies and totalitarian regimes alike.

As if to secure these illiberal practices, world leaders along with the new tech-giants have launched the greatest practice, yet, of neo-liberal capitalism, what author and scholar Shoshana Zuboff has called Surveillance Capitalism[23] made possible by the exploitation, if not colonisation of the internet. The inter-marriage of the digital with neo-liberalism has created the greatest and most disturbing threat, to date, to the concept of democracy because not only is it massively, globally tentacular, it is also invisible and unseizable; and by its virtual nature, ultimately un-legislatable – impossible to regulate. Indeed, those in control of this system, the tech-giants themselves claim that it is too complex as a technology to legislate – what a self-serving let-off, if there ever was one! And lest we believe we can trust governments to limit the abuse of this technology, it is worth recalling that in China all aspects of digital surveillance are used by Xi Jinping's state machinery to assert total control; thus, facial recognition is used everywhere in China; surveillance of consumption is easily practised since the great majority of purchasing is done on mobile phones; and once China's 5G is fully installed at home and abroad, there will be no limits to this control.

I'll conclude this section on Dirty Capitalism with a brief discussion of the effects of internet and social media (essentially a deregulated system) on the principles of democracy. For, these too are integral to the harming of the biosphere and so their impact on the social foundation of humanity cannot be underestimated. We are faced with a new kind of colonisation, argues Zuboff, this time of our mind-life. Once you have entered the internet cage, or to give it a nice term, the digital domain, invisible colonisers (the iCloud, Google, Facebook being

---

21 Ma Jian quoted in interview by Claire Armistead, 'Freedom can't be taken for granted. We have to remain constantly vigilant' (*The Guardian Review Section*, 03 November 2018, 22) (full page reference 20–3). See also BBC Two's recent three-part documentary series on XI Jinping *China: A New World Order* 5–26 September 2019.
22 Ma Jian, 'Surviving Tiananmen' (*The Guardian Review Section*, 01 June 2019, 37).
23 See: *The Age of Surveillance Capitalism: The Fight for a Human Future at the New Frontier of Power*, London, Profile Books, 2019.

obvious examples) appropriate our experiences, our innermost secrets and thoughts even, just by dint of our using their services, and then trade them for profit. Google, under the guise of organising all human knowledge, in effect controls it. As we search this powerful engine for the information we seek we, in turn, are searched ourselves. Just as before, when we entered the robotic age in the late 1960s, Baudrillard spoke to us about the de-responsibilisation effects of *that* new technology (that is, by replacing man's labour with robots, they became the system of objects that could fail, rather than man), here too we are faced with a new man-made technology for which those who put it in motion (the tech-giants) refuse any sense of social accountability. Instead, they build their fiscal empires on publicly funded data[24] *and* the details of our private lives by selling on the accumulated data to markets, advertisers, and campaign organisers (to name but the obvious groups).[25] Under the guise of our accessing more information, that information then serves as a tool to predict our behaviour and exploit it to the advantage of whichever body purchases it. One could also call it Platform Capitalism, virtual, invisible, intangible as it is. Surveillance Capitalism of this order is profoundly undemocratic. Knowing our properties, it is far easier to predict our choices, tastes, and behaviour and thereby manipulate them. Processing the accumulated data into algorithms allows for precise targeting of individuals; thus making the selling of politics and ideologies, for example, a far more accurate affair (as we have seen recently in a number of referenda and elections). We have entered the era of hacking humans; an inevitable outcome, one supposes, of digital technology at its worst.

Scientific thinkers like James Lovelock (and the vast majority of A.I. scientists, doubtless) hold the view that the Anthropocene is a consequence of life on Earth and is a product of evolution which will soon be superceded by what Lovelock calls the Novocene, an age of hyper-intelligence in which A.I. and computer technology will surpass the human in all things, and in which our role will be to keep the planet

---

24 That is, the information that public bodies produce. Public sector information includes government data, geographical information, statistics, weather, data from publicly funded research institutions, digitised books from libraries (etc.).
25 The defunct Cambridge Analytica is not the only lobbying company to affect recent electoral outcomes; Lynton Crosby's CTF Partners ran 'pages on behalf of the Saudi Government, as well as political campaigns on behalf of major polluters and big-budget campaigns pushing a no-deal Brexit.' See Jim Waterson, 'Lobbying MPs want to question Crosby on propaganda' (*The Guardian*, 03 August 2019, 12). Both lobbying groups have close ties with Facebook (and of course politicians!).

at a habitable temperature because the machines will need organic life in order to survive.[26] Deep ironies here, wouldn't you say, that we are forced by our own machinic-creations to maintain a liveable habitat (something we currently seem incapable of doing)?

Wherever our technological «prowess» may lead us, none of it alters the fact that Nature is the ultimate carbon capture and storage machine; that before humans came along nature did a very good job of regulating the climate in the warm ages (the inter-glacials) and the ice ages, as well as the transition in between.[27] It is visibly the case, also, that unless we rethink our place in time, our relationship with Earth's deep time, as opposed to our current notion of time as speed and unlimited growth, there is very little that the planet can have in store for us.[28] Every year, we use up the planet's natural ability to resource our needs by leaps and bounds – in 2019 the Earth Overshoot Day was July 29. Here, Figure 2.1 shows a graph tracing the folly of our grasping madness. Its start date of 1971 heralds the first acknowledgement that, what was then known as the 'greenhouse effect,' was seriously damaging our ecology.[29]

If we accept the premise that the story of modern democracy is also the story of modern capitalism – as many Neo-Liberalist economists would argue – then, we come up against a startlingly obvious contradiction in terms, because the stark evidence is there, that economic divides still prevail. Democratic societies contain 'vast inequalities of power and education' (as a visual iteration of this, see any of Ken Loach's or the Dardenne brothers' social-realist films); and 'their media have always been driven by partisan and commercial imperatives' – indeed 'ever since its origins in the late 18th century, modern democracy has had a peculiar

---

26 James Lovelock, *Novacene: The Coming Age of Hyperintelligence*, London, Allen Lane, 2019. I should add that Lovelock is a Gaia Theorist; his views expressed in *Novacene* are to be read as a warning as much as a prediction.
27 See Graig Bennett's Editorial 'Nature is the ultimate carbon capture and storage machine,' In: *Earthmatters*, Spring 2019, 9.
28 See Robert Macfarlane's essay 'Up from the depths,' *The Guardian* (Review section), 20 April 2019, 8–13; and his book *Underland*, London, Hamish Hamilton, 2019.
29 It is noteworthy that scientists in the 1960s were already talking about the effects of the levels of carbon dioxide in the atmosphere and that Rachel Carson's *The Silent Spring* (1962) was a fierce warning of the effects of pesticides on the environment. A recent French film relates the disastrous effects of pesticides on farmers: *Au nom de la terre*, Edouard Bergeon (2019).

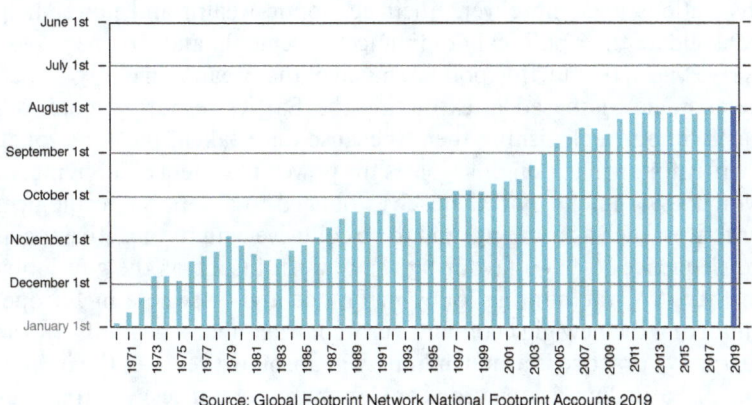

Figure 2.1 Graph of earth overshoot day 1971–2019.[30]

relationship to truth.'[31] In their book, *Why Nations Fail*, Daron Acemoglu and James Robinson set out a clear reason as to why nations, ultimately, fail and why democracies do not really work towards the principle of equality. To begin with, the central flaw lies in the synergistic relationship between economic and political institutions (2013: 81). Arguing that there are two types of political institutions, the inclusive and the extractive, wherein the former endeavours to foster economic growth (for the better of society), and the latter impedes and blocks growth (and keeps both the nation and the citizenry poor), they then add that, whilst 'the politics of some societies lead to inclusive institutions that foster economic growth,' for the most part 'the politics of the vast majority of societies throughout history has led, and still leads today, to extractive institutions that hamper economic growth' (83). Less a question of failure, more a fact of deliberate intention. Indeed, Naomi Klein had already alerted us in her 2007 book, *The Shock Doctrine: The Rise of Disaster Capitalism*, that the social breakdowns witnessed during decades of neo-liberal economic policies are not accidental, but in fact integral to the functioning of the free market, which relies on disaster and human suffering to function!

---

30 Sourced from www.overshootday.org (accessed 01 August 2019).
31 Fara Dabhoiwala 'Book of the Week Review of *Democracy and Truth: A Short History*, Sophia Rosenfeld' (*The Guardian Review Section*, 22 December 2018, 14–15).

Predominantly, political elites select economic models and institutions that consolidate their political dominance. Thus, in the West, the capitalist growth model enriches these same elites and economic institutions and, moreover, 'their economic wealth and power help consolidate their political dominance' (Acemoglu and Robinson, 83). As we know, the starting point to much of that wealth, in our so-called democracies, is based in extracting the Earth's resources (coal, oil, gas) without replenishing them (because once taken, they are gone). Swiftly following upon this, comes the powerful system of lobbying, in which governments find themselves obliged to accept bad ecological practices (such as fracking) and to be collusive with the negative social consequences of laws favouring the powerful (such as the gun lobby in the USA, the hunting lobby in France and elsewhere in Europe, endorsing the new gig economy, zero-contract hours).[32] Similarly, the extractive practice is as much in play in absolutist regimes (be it African or Latin American nations or Communist countries). All that separates the two politico-economic systems is the *degree* of distribution. In Western democracies there is some form – however pitiful in many circumstances – of economic distribution (through the levying of taxes into a governmental budget that is then distributed to various services, including social services). In absolutist regimes, the system of capital levy – whether selling off valuable minerals or exploitation rights to economic elites or foreign bodies, or «mismanaging» fiscal revenue – seldom brings about the distribution of the revenue much beyond the pockets of the rulers and their cronies. So, no matter whether we are speaking of a rich or a poor nation, economic well-being comes down to the way in which political power is exercised – and, in the main, this power is mostly monopolised by a narrow elite, organising society for its own benefit. A government that does not tax and distribute fairly is guilty of institutional corruption. The consequences are political in that democracy is not safeguarded. Indeed, what we are left with is a lack of equality and a lack of political rights. The idea that state and capital can live apart, which is an ideology of deregulation, is of course

---

32 As Houman Barekat rightly points out 'Job insecurity and wage inequality have been rising ever since the years of Reagan and Thatcher, but the advent of digital technology has exacerbated this trend, making it easier for such companies as Uber to assemble and manage large armies of low-paid contract employees. (...) "gig economy" work invariably means no health insurance, no retirement pension (...) an economy built on insecurity is a public health time-bomb.' 'Book of the week review: *Lab Rats: Why Modern Work Makes People Miserable*, Dan Lyons' (*The Guardian Review Section*, 29 December 2018, 12–13).

a nonsense. 'The modern state and economy are twins born together in the seventeenth century,' since when their relationship has 'matured into one of ever-greater interdependence.'[33] And yet, that interdependence is constantly assailed. To the point, as indicated above, that when economic and political interests combine, what gets overlooked are the questions of state, such as social well-being and welfare. Frankly, the only time in recent history when state and capital have been held in balance was the post-war era of State Welfare as practised in Britain and France in particular.

2.3 Anthropocenic madness: deregulated states of mind, capitalism, and its consequences – three classical Hollywood movies seen through the Doughnut

Let us now consider three mainstream American films – *Mildred Pierce* (Michael Curtiz, 1945), *Tulsa* (Stuart Heisler, 1949), and *Giant* (George Stevens, 1956). All three are predicated on the three-act arc narrative and, as we shall see, are emblematic of this deregulated state of mind, which surely approaches a kind of madness. When looked at through the prism of the Doughnut, however, a deeper picture, still, emerges. *Mildred Pierce*, often considered a film noir, is also a film about the capitalist drive for wealth and status. Based on the very American meritocratic belief that any individual can make their fortune by dint of hard work, the narrative follows the trajectory of a self-made woman whose absolute certainty that wealth means happiness (in particular her daughter's) ultimately leads to murder. *Tulsa* and *Giant* are two films with at their heart an obsession with gaining wealth at all costs, this time within the domain of oil-drilling. In the case of these last two films, some apparent awareness –the one environmental (*Tulsa*), the other social (*Giant*) – prevails.

Mildred Pierce is set in the first half of the 1940s and not, as in the original novel, the period of the Great Depression (this change of era will be of significance, see below). Mildred's fortunes have sunk as a result of her husband, Bert's business failure. An embodiment of middle-class America, the eponymous heroine's first drive is to maintain her family's status quo. To achieve this, she successfully sets up a series of restaurants (having first started work as a waitress in a diner).

---

33 See Adam Tooze's Review of Donald Sassoon's *The Anxious Triumph: A Global History of Capitalism* (2019, Allen Lane) (*The Guardian Review Section* 03 September 2019, 16–17).

Her rapacious drive to succeed leads to her marriage breaking up. But Mildred is not alone in extreme ambition. Her equally class-conscious and over-ambitious daughter, Veda, is also a major factor behind Mildred's drive to make money. However, Veda has nothing but disdain for her mother's source of wealth (basically, trade) and she pursues an advantageous marriage to a young man of social standing whom she subsequently attempts to blackmail. When that fails, Veda pursues a career as a night-club singer (much to her mother's horror). Totally distraught, Mildred seeks an advantageous marriage for herself that will satisfy her daughter's social-climbing desires. So she marries Monte, a feckless society playboy. Monte's greater prestige and trappings of wealth (a beach house on the California coast) attract Veda back into the fold to the point of having sexual relations with him.

By now, all those surrounding Mildred (with the exception of her first husband, Bert) have conspired in different ways to ruin her business. She goes to the beach house to confront Monte who has cheated her out of her money. But, there, she finds Veda in his arms. In a showdown between the two women, Veda declares that Monte is leaving her mother to marry her. Monte rejects Veda, Mildred runs out of the house, and Veda shoots Monte dead. Cajoled by Veda, Mildred attempts to pin the blame for the murder onto her former financial backer, the rather seedy and unprincipled Wally. Meantime, her first husband, Bert, has confessed to the murder to protect Mildred and (presumably) Veda. But the all-knowing police (to whom this sorrowful story of greedy ambition has been narrated in the form of flashbacks) are fully aware that the culprit is Veda and she is led away in handcuffs – still rejecting her mother. Mildred leaves the police precinct with Bert (one assumes she will now remarry him and who knows, together they may start up the business all over again!).

The relocating, in time, of the film's narrative from the Great Depression to the war-years takes the story away from the globally important economic disaster caused by stock-exchange speculation and reconfigures it into a more socially economic one: a war period during which women did go out to work only to find that, at the end of the war, they were expected to return into the domestic sphere. But, in fact, elements of the Great Depression do linger on in the film. For, Mildred's motivation for money is linked purely to her husband's loss of fortune. Furthermore, there is no mention of the war, and none of the men immediately in her entourage are sent off to fight. Instead, Monte is content to live off Mildred's labour; Wally is content to live off the work of his night-club performers which include Veda. Only Bert seems capable of decent behaviour. So what we are confronted with here is a layering of exploitation and relations of power, more so than the more feminist reading often made of Mildred's plight of a working woman returned to domesticity.[34] Exploitation works both ways: Mildred, knowing Wally is attracted to her, is perfectly happy to exploit her friendship with him to get the necessary money to buy her first restaurant; she uses Monte's social standing to lure her daughter back home. In terms of power relations, all four (Mildred, Veda, Monte, and Wally), at different moments within the film, are in ascendency. What is evident is that wealth-acquisition is the focal point of all four – and they'll stop at nothing to get it.

The other significant distortion from the original text is the introduction of Monte's murder as the dénouement. This was imposed by the Censorship (Hays) Code which required (in an attempt to control the glorification of villains, particularly in thrillers and gangster films) that evil-doers be punished for their crimes. This is somewhat bizarre given that, without the imposed murder, effectively no crime has been committed, that is merely social climbing, some financial dishonesty (but all within the family) and moral misdemeanours (with the exception of Veda's attempted blackmail, which fails). But no, the Code will have it: one man (Monte) and one woman (Veda) must perish for their evil-doing. The other two (Mildred and Wally) are issued with cautionary slaps on the wrist.

These two departures from the original text allow us already to see – in relation to Raworth's Doughnut – that not just *gender inequality* is at play, but how complex the mechanics of it are and the *lengths* to which patriarchy must go to enforce it. If indeed the film's message is, *one*, that women's economic independence must be curtailed and their greedy social ambitions brought to a quick closure, then I would point out that, *two*, this does not occur until Monte has managed to purloin Mildred's hard won riches – which then obliges her, *three*, to reconsider her position (maybe Bert wasn't so bad after all!). In a sense, therefore, patriarchy has a triple bite of the cherry. I would also add that Mildred's business common sense is represented as considerably more adept than Bert's – so she is a woman in a man's world doing better than the men. A kind of masquerade is at play, therefore – woman-as-man. In gender terms, Mildred is in excess by dint of her masculinity, rather than any hyper-femininity (a woman in

---

34 See Annette Kuhn (1982) *Women's Pictures: Feminism and Cinema*, London, Boston, Melbourne, and Henley, Routledge & Kegan Paul, 29–35. But it is clear that by relocating the film into the war/post-war period the narrative does indeed lend itself to such an interpretation.

excess of her femaleness, which is what is more readily associated with the term female masquerade). And it is for that reason, too, that she is «punished». First, she is brought down (to her place) by Monte's underhand scheming; second, she loses all her trappings of motherhood (Veda is lost to her and so too her other, neglected younger daughter, who died whilst Mildred was away making her fortune). In this light, Mildred is a bad woman or non-woman (in playing the male capitalist game) and a neglectful mother.

If we further consider the dynamics of the film, a number of other transgressions of the *social boundary* emerge. Both women, as if mirroring each other, are driven by a grasping need for money and prestige no matter the cost to others, so can hardly be considered as being invested in *social equity*. This overwhelming urge for getting rich, as the be-all and end-all of a life's ambition, merely reproduces the patriarchal model as embodied by Wally and Monte. Thus, in this context, *gender equality* takes on a rather distasteful tenor here. Although, of course, by the end of the film both women are brought to order once more under the patriarchal law (literally for Veda) – an outcome that does nothing to resolve gender inequality, I fear.[35] In terms of *social equity*, both women's sense of entitlement ties them into the capitalist system of class hierarchies – a sense that they deserve to be in and belong to the wealth-classes. Their aspirations to social status leads them, as with Veda, to disdain trade as a source of wealth, or, as with Mildred, to be ashamed of it – until, that is, she becomes the successful businesswoman («meriting», therefore, her wealth). On several levels, the two women also transgress codes of *justice*; Veda for her blackmailing attempts, her quasi-incestuous relationship with her stepfather and for trying to cover up her murder; Mildred for attempting to put the blame for the murder on an innocent man Wally (and justifying it to herself because of his sleaze) – all to save face and her daughter. In Doughnut terms, then, we can see how the film stands as a metaphor for what is lost in the pursuit of Capital: social equity and justice.

Turning now to our two «oil-drilling» films *Tulsa* and *Giant*. These are classified generically as Westerns, interestingly; although perhaps this makes sense, since the land upon which these narratives evolve is indeed land mostly taken (stolen) from the Indians as part of the great

---

35 And for Mildred, it is the other rule of law of the father, the spoken word that prevails ultimately. For her version of the story (told in two flashbacks), is finally corrected by the police detective (emblematic of the law) whose third and final version (again in flashback) establishes the true story (the narrative is safely back in the hands of the men, therefore).

drive West by prospectors in the second half of the 19th century. Let us begin with *Tulsa*, which is set in the 1920s, the period of great expansion in the oil-drilling business, particularly in the Tulsa area of Oklahoma making Tulsa the 'Oil Capital of the World.'[36] Once the first oil strike was made in 1905, at Glenn Pool (just outside Tulsa), oil prospectors, known as Wildcatters flocked to the area to lease land from farm owners – gambling that the lots they leased would bring them untold riches. As we learn from *Tulsa*, some of the farmers leasing out their land were Indian land owners (those who had a small tract of their land returned to them, under the Dawes Severalty Act).[37] Unfortunately, by the time of the 1920s, the farming industry was struggling.[38] So, given that an average farm at that time was between 100 and 170 acres (although for Indians it would have been more like 10–20 acres[39]) and that leases to drill for oil were sold at $2 per acre, with one-eighth interest in any oil-production, farmers found this trading of their land an answer to their prayers.[40]

Tulsa opens with a set of panoramic shots of the Oklahoma landscape, including images of cattle-grazing territory, accompanied by a voice-over informing us that this land, which was formerly Cherokee country, was an 'ideal land' but not for what it was (that is, a natural open landscape), but for what it could become: 'the oil underneath, it had to come out,' we are told. This naturalising discourse establishes the idea of the inevitability of oil-drilling, of growth, without

---

36 More money was made on the Glenn Pool oilfield than the California gold rush and Colorado silver rush combined. https://aoghs.org/petroleum-pioneers/making-tulsa-oil-capital/ (accessed 02 September 2019).
37 Land settlements for Indians were established by the Dawes Severalty Act 1887. It was a case of allotting a small portion of what was really Indian land (but known by the 1850s onwards as Unassigned Lands, the Indians having been reassigned to Reservations) back to Indians in the light of the impending land-rush of 1889 for Oklahoma Indian Territory. Under the guise of protecting Indian rights to some land, this Act allowed millions of acres of original tribal land to be consolidated into tracts of farmland and sold to crop farmers (corn, cotton, etc.). https://newsmaven.io/indiancountrytoday/archive/native-history-land-rush-for-oklahoma-indian-territory-begins-zyaP2-UBRE2EQitF83mnrw/ (accessed 02 September 2019).
38 See www.okhistory.org/publications/enc/entry.php?entry=FA019 (accessed 02 September 2019).
39 See www.archives.gov/files/education/lessons/fed-indian-policy/images/application-page-05.gif (accessed 12 September 2019).
40 The Mineral Leasing Act of 1920 set this rate for petroleum (of course Wildcatters may have offered different prices to struggling poor farmers; in 1905 the price per acre was 3 cents…). https://en.wikipedia.org/wiki/Mineral_Leasing_Act_of_1920#Petroleum (accessed 02 September 2019).

so much as a by-or-leave for the consequences to wildlife, landscape, a former way of life of, first, the Indians and latter the cattle-ranchers. Indeed, as early as the opening sequences, any talk of the value of this 'ideal land' is reduced to a binary opposition between oil-drilling and cattle-raising, thus overlooking the territory's earlier Indian heritage – a bleaching out of the Indian historic relationship with the land (which was one of co-existence, not ownership).

As if to make this binary totally explicit, what triggers the story we are about to follow and which draws it to a conclusion, is the death of cattle, due to oil deposits polluting the creek's water. By the very end of the film, the war between cattle and oil is brought to some kind of resolution. The oil-magnates (or Wildcatters) accept the conservationists' argument that there must be an equilibrium between the two. An agreement is drawn up limiting oil-drilling to one rig per ten acres, with the intervening spaces to be cordoned off for cattle-rearing. No mention, however, is made as to how oil pollution of the water will be contained; and the resultant landscape remains denaturalised. The film closes with images of the former beautiful territory colonised by a mass of oil rigs and enclosures for the cattle. Any trace of the Indian heritage is itself completely erased.

Central in all of this is the trajectory of Cherokee. She is the daughter of a cattle-rancher and is a quarter Indian. The death of her father as the result of an argument with the nearby oil-drilling company, owned by Tanner, leads Cherokee to confront Tanner and demand compensation (her father was killed by falling débris from an oil gush). Tanner refuses, adding 'take me to court if you want, I've got plenty of money so you'll lose' (reminding us ineluctably of the practice of giant corporations who are able to hold off forever paying out any compensation). A chance encounter with another oil man, John Brady, leaves Cherokee with leases to drill on farmland near her home. As success in drilling follows success, she becomes extremely rich. Nothing matters more to her than beating Tanner at his own game. That is, until, intent on becoming richer still, she joins ranks with him with a view to the full exploitation of her own farmland. In so doing, she betrays her childhood friend, a fellow Cherokee, Jim to whom, when her father died, she had given the farmland so that he could continue with her father's cattle-raising. In exchange for this gift, he had leased to her *limited right*s to drill for oil. She is now set to break that understanding. With her new business partner, Tanner, she takes Jim before the judge to fight her corner. The judge treats Jim abusively ('I question the man's competency') insinuating that the 'Indian must be crazy' not to see the necessity to drill this oil out and threatens to have him institutionalised. Cherokee does not defend her old friend, and he rushes out in despair. The film concludes with Jim, back on the farmland, discovering dead cattle in the creek. Appalled, he sets fire to the oil-polluted water (just as Cherokee's father had done at the beginning of the film). But this time the fire catches with a fury and spreads to the oil rigs. A conflagration ensues. Once the fires are brought under control, Cherokee comes to her senses ('she's back from the mountains') and agrees to the conservationists' treaty, as does Tanner.

Gender equality comes to the fore in pretty much the same guise as it does in *Mildred Pierce*. Cherokee is another strong-minded woman who is determined to show she is as good as the men – even to the point of showing them in a dice-throwing game, at the local casino, how she is an expert at winning. Curiously, this is the one time that she invokes her Indian heritage. Claiming to possess a knowingness the Whites don't have, she uses lucky chants to tell the dice what to do (and they do). But, as with Mildred Pierce, playing the men's game leads to her overshooting in the end (the conflagration mentioned above, being the most obvious example). In fact, the dice game is a first warning which she neglects to heed. For she does lose on her final throw. Tanner challenges her to a huge bet. She undershoots – her Indian-ness forsakes her! At the film's end, however, unlike Mildred before her, Cherokee's driving ambition is chastened rather than punished – she accepts compromise; marries her devoted conservationist (Brad); and, begging Jim's forgiveness, restores her friendship with him.

Tulsa sets up a deliberate binary between pollution and conservation.[41] In so doing, its scriptwriters offer us an interesting cautionary tale on capital growth and the obsessive pursuit of wealth (unusual perhaps for 1949, when the USA had become *the* dominant super-power). Reading the film through the prism of Raworth's Doughnut, we can determine a causal chain of transgressions on the planetary sphere: land conversion to oil-drilling, causing chemical pollution, leading to

---

41 I am tempted to add that the carbon footprint for this film must have been quite significant, especially the huge conflagration at the end (which doubtless explains why the film was nominated for an Oscar for Special Effects, in 1950). Here's what the reviewer, Bosley Crowther, writing in 1949, has to say:

> He (Stuart Heisler) had a whole forest of derricks, oil tanks and equipment set ablaze in as fiery a Technicolored burn-out as ever you're likely to see. And he had Mr. Preston, Miss Hayward and, indeed, the whole gol-darned cast dash into the conflagration with wild expressions, bull-dozers and dynamite.
> www.nytimes.com/1949/05/27/archives/tulsa-story-about-oil-fields-with-susan-hayward-new-feature-at-the.html

fresh water pollution. The lack of regard for the proper development of the land (as advocated by the conservationists) produces a wasteland where grass cannot grow and cattle cannot graze. Oil overspill leaches into the water rendering it noxious and fatal to those who drink from it.

But it is the insidious cocktail of calamitous conduct (gambling, obsession with growth and wealth) and the ensuing environmental abuse which that behaviour brings about and its social consequences that is most clearly exposed by mapping through the Doughnut. The inter-relatedness of environmental abuse and social inequity is made transparent not just through the judge's racist attack on Jim, but it is also there in Tanner who will over-ride everyone, including Jim and other Indian farm owners, to get what he wants. Worse still, it is present in Cherokee, whose obsessive compulsion for more leads her to drag one of her own heritage, literally, down into the earth-ground she has destroyed through her greed. In short, she pollutes the very land of her ancestors, the Cherokees. She sacrifices her moral compass, so strongly asserted at the beginning of the film, to the ventures of growth capitalism at all costs – at all deregulated costs since she will stoop to anything to have her way, pushing the boundaries of the law beyond their remit. Wealth creation of this magnitude is a madness that brings in its wake a total disregard for *social justice* and *equity* – in such conditions, the social sphere is not just in shortfall, clearly it is in freefall.

And so onto *Giant*. If *Tulsa* is inadvertently an early example of films exposing the impact of extractive economies (in the form of oil-drilling) on the environment and its on-going effects on human consciousness, *Giant* is determinedly clear on its starting point: racism as it emerges from within settler-states, themselves often a product of land-grabbing and a sense of land entitlement. The film is an adaptation of Edna Ferber's 1952 novel of the same name. A victim herself of anti-Semitic abuse as a child and young woman, Ferber wrote against racial and cultural discrimination in several novels, *Show Boat*, 1926, and *Cimarron*, 1929 (this latter novel set at the time of the Oklahoma Land Rush). *Giant* was her third one to deal with racism. And as with her earlier novels, some of her central characters are inspired by real events or people. Jordan 'Bick' Benedict II is based on Bob J. Kleberg Jnr, the wealthy cattle-rancher of the King Ranch in Southern Texas, and Jett Rink on the Wildcatter oil tycoon Glenn Herbert McCarthy.

Texas is the front and backdrop to the film's epic narrative. Texas, formerly Comanche land, was first taken by the Mexicans in the early 19th century, then by the Americans in 1848, after the Mexican-American

War. By the mid-19th century, wave upon wave of European immigrants (Germans and Czech primarily) had arrived and settled in Texas, as did Americans, and, later, freed black slaves. Meantime, from the 1850s onwards, the Comanches not only lost their territories, their numbers were decimated by European diseases (Small Pox and Measles) and, by 1875, they were removed to a reservation in Oklahoma. By the 1880s, large cattle-ranches were established, made possible by sales that allowed individuals to accumulate massive stretches of land (up to half a million acres or more) which they then enclosed with barbed wire. As for the Mexicans, those who remained – known as Tejanos (many of whom were formerly poor land-croppers) – became indentured to the American cattle-ranchers. Their status, by the time of this film's narrative (1920–50), was the lowest of the low, and to that effect social segregation was enforced against them. In short, as a state, Texas was a messy affair of ownership and land-grabbing, of classism and racism – something which the heroine of the film, Leslie points out at the very beginning of *Giant*.

Here is a brief synopsis of the film. Bick is the third generation of the wealthy Benedict cattle-ranching dynasty and owns half a million acres of Southern Texas, an extremely barren and arid part of the state. He marries a wealthy socialite, Leslie Lynnton, who hails from the 'East,' namely, Maryland, a green and lush land where her father raises horses. The gritty down to earth Westerner marries the refined, courteous, socially conscious Easterner – the first in a series of oppositions that underpin the narrative – and brings her to live on his ranch. Bick's sister, Luz, is as unen*light*ened as her brother in her contempt for the Tejanos they employ on the ranch (both utter sentences such as: 'those people,' 'lazy good for nothings,' 'migratory mess'). She is culturally dismissive of Leslie, 'you lot from the East you're not tough,' a feeling shared by her brother it would appear, and one which Leslie quickly refutes ('I'm tougher than you think') and later exemplifies through her conduct (helping the Tejanos obtain medical care, standing up to Bick's misogyny). Also on this huge ranch is an odd-job's man, Jett Rink who for two-thirds of the film is mostly on the periphery, peering in and observing this wealthy family. He is a drunk and as much of a bigot as Bick (calling Tejanos 'wetbacks'). Bick's loathing for Jett is matched by Luz's affection for this misfit, and, upon her death, she leaves him a tract of land on Bick's estate which Jett immediately encircles with barbed-wire fencing. Jett eventually strikes oil and becomes extremely wealthy.

During this whole period, Bick and Leslie have produced three children and Jett maintains an unrequited passion for Leslie. Once the children are adults, it transpires none of them want to take over Bick's ranch and he leases the land to Jett so he can expand his oil-drilling (rather than giving 'it back to those dirty old Comanches' as he puts it). Bick's son, Jordy, becomes a doctor; marries a Tejana, Juana-Maria; and works in the Tejano villages to help the poor. As the film draws to a close, Jett holds a huge party in his own honour. By now, his former drunkenness has become full-blown alcoholism. Bick's family are in attendance. Jett racially abuses Juana-Maria. Fights ensue: Jordy loses against Jett in his defence of his wife; Bick has it out with Jett for insulting his son. The Bick family leave. Stopping at a diner on their long drive home, once again Juana-Maria and her baby son are the targets of racial abuse, this time from the diner's owner. Bick tries to defuse the situation by reminding the owner who he is. But it is only when a Tejano family enter the diner and are refused service that Bick finally awakens to the owner's racism and fights back. He loses the fight, but as Leslie tells him, he won the moral battle: 'in that fight, you became my hero.' For once, he stood up for the oppressed.

The desolate, wind-swept landscape of the Texan cattle-ranch seems a perfect foil to harsh goings-on within the narrative of this film. As indeed does Bick's home, a Second-Empire Victorian Mansion stuck in the middle of nowhere. Not only is the architecture completely out of context to the environment (to all appearances it looks more like an Addams Family Victorian Gothic horror house), its pretentiousness at taste tells the viewer a great deal about Bick's family heritage. In meritocratic terms the Benedict ranch is King – the mansion stands as if to assert this. However, as Jett astutely remarks, as an aside early on in the film, 'ain't nobody King in the country.' The savage, windy territory upon which this house stands provides a grotesque visibilisation of wealth – but wealth in a void space, where nothing grows, where all is dust, where, at the end of the film, a huge twister blows through (doubtless off the Gulf of Mexico) as if to remind Bick and Jett of the precarity of their ownership. This harsh space of emptiness finds painful echoes in the noise produced by what occupies the landscape, such as the lowing of the huge numbers of cattle, their cries of despair as they are lassoed and branded, the metallic clinking of spurs, and, later, the metronomic noise of the oil rigs. This is a place of anthropocenic violence, a fomenting ground for hatred (Luz towards Leslie, Bick towards Jett), racism (Luz, Bick, Jett), violence (Luz torturing Leslie's horse, the numerous fights), disregard for the poor (Luz, Bick, the other wealthy ranchers). Anthropocenic violence as represented in this film is a direct outcome of ownership. We note, for instance, how

Jett stakes out his newly inherited land with *barbed-wire* fencing. But this violence is manifest at all levels, starting at the familial, as seen in Bick's bullying ways towards his children (particularly his son), and spreading all the way through the different social hierarchies. Only the Tejanos, with the exception of Bick's foreman Gomez, have the resigned wisdom of the oppressed.

This grotesque visibilisation of wealth and violence is matched to the point of total excess in the form of Jett's developments (not just the oil rigs, but other trappings of wealth, airplanes, huge emporia, luxury hotels) and his violent outbursts and alcoholism. Jett is the brute-force mirroring of Bick (whose veneer of class barely disguises his ill-adjusted temperament). Bick, his sister Luz, and Jett are different manifestations of the kind of madness that a deregulated state of mind can produce. Bick and Luz are stuck in a deeply engrained belief of continuity ('this is the way things are and things are done') which is an outcrop of their sense of ownership – or their insecurity about their rights of ownership which is why they doubly declaim their authority (a vestige, one could surmise, of colonialism). Jett is crippled by resentment at having nothing, to the point of embodying that nothingness (he slobs around the ranch in a permanent state of inactivity). But, like Bick and Luz, he upholds the same racist and bigoted attitudes of the wealthy White ranch-owners, and as such endorses a world-view that disadvantages the impoverished Tejanos. So it is small surprise, therefore, that – once his lucky inheritance-break brings him huge wealth – he becomes a caricatural version of Bick. And, as if to echo the caricature, the savage arid plains of before are now littered with the oil-rigs at work – literally, we observe them, at the film's end, through the front door of Bick's mansion.[42]

As a trio of personages, then, we can see how a perpetuation of beliefs and behaviours has an impact on both the ecological and the social boundaries Raworth speaks of. Balancing this impact, however, are the forms of resistance, starting with Leslie's sense of *social justice* and defiant assertion of *gender equality*. Hers is a voice that reminds the others of the falsehoods behind their frontiersmanship (the right of White settlers to seize whatever land they wanted): 'We stole Texas, didn't we, from Mexico?' she asks. In that 'We' she does not, we note,

---

42 Even if the two grand-children are in the foreground, that too doesn't bring much hope, especially since Bick – despite being slightly more enlightened around questions of racism – still insists on referring to his grandson as a 'wetback.'

exclude herself from that injustice. Rather, she recognises what has been done in her name. Moreover, as she witnesses the racism and bigotry exacted upon the Tejanos, she acknowledges how this prerogative of taking has built an ownership mentality that is inured to the brutality it exercises in its wake. Through her acts of kindness towards the Tejanos she repeatedly questions the quite savage assumption that to own is to control everything and everyone ('including me?' she asks Bick at one point). Just as she fights for the health and well-being of the poor, similarly, she argues the case for *gender equality*. She reminds the men around her that: 'I have a mind of my own'; and when Bick tries to exclude her from a political discussion he is having with fellow ranchers ('Leslie, this is men's stuff'), she declares loud and clear: 'your problems date back 11,000 years, which is to masculinise politics so that women can't understand.' Unsuccessfully as it transpires, since she refuses to be pushed aside and understands full well what they are talking about.

Leslie's social consciousness is echoed in the behaviour of her twins Jordy and Judy both of whom refuse to embrace the mentality of prerogative ownership. Jordy elects to be a doctor and uses his skills to help the Tejanos towards a better quality of life; Judy turns down her father's offer of the ranch, declaring she and her husband want to establish a small ranch. The brother fights for *social justice*, the daughter for *sustainable growth*: the safe place for humanity, which the biosphere of Raworth's Doughnut sets out.

Here then is a diagram which locates all three films in relation to the ecological and social boundaries of the Doughnut (Figure 2.2).

The diagram below provides a pretty stark representation of the Anthropocene at work. *Tulsa* and *Giant*'s commercial practices so evidently transgress the ecological ceiling. Practices that begin with the historical land grab, swiftly followed by land conversion first to cattle-ranching then to extractive oil-drilling, which, in turn, leads to chemical and water pollution and biodiversity loss. Nothing in their business enterprise enables the safe space for humanity. The impact of these transgressions can be clearly seen on the social ceiling, where many of the most essential elements for humanity's well-being are in shortfall. As for *Mildred Pierce*, in terms of social equity, the picture is equally bleak. Indeed, where all three films are concerned, social equity and justice and the pursuit of sustainable growth on smaller ranches are values asserted in the behaviour of a socially conscious few – all to be found in *Giant*: through Leslie, Jordy, and Judy. Everything else points to continuity in the same set of beliefs of growth economics. When viewed through the prism of the Doughnut Theory, the idea that the three-act arc narrative

Film and the Anthropocene 45

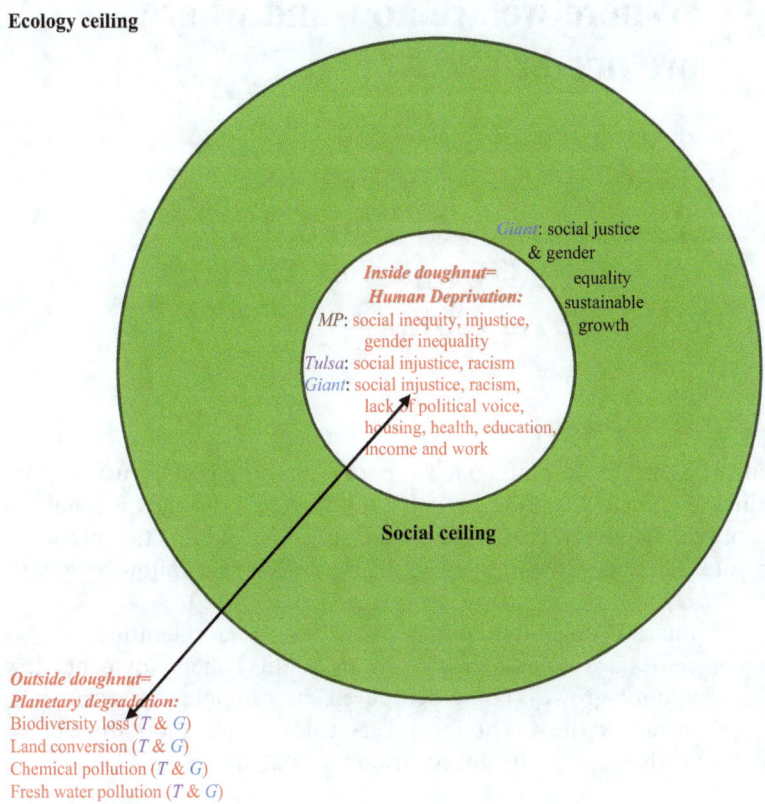

Figure 2.2 Locating *Mildred Pierce (MP)*, *Tulsa (T)*, and *Giant (G)* on the Doughnut.

of mainstream cinema brings resolution is hardly a satisfactory one. If anything, it exposes the delusional nature of such narratives. But it also points to what needs to be more widely (econometrically) acknowledged to effect change. And that's the topic of the next chapter where we will examine three mainstream film narratives that stand as expository examples of the madness of the Capitalocene at work.

3 Where we are now and where we need to be

From source and sink to take→make→use→re-use – *Erin Brockovich* (2000), *Wall Street* (1987), and *The Wolf of Wall Street* (2013)

3.1 Source and sink

In 2018, the UN commissioned a report, submitted in August 2019, on climate change. The resounding result was that capitalism is not the economic model to prevent climate change, indeed it is the direct opposite. The scientists drawing up the report have the following to say:

> Climate change and species extinction are accelerating even as societies are experiencing rising inequality, unemployment, slow economic growth, rising debt levels, and impotent governments. Contrary to the way policymakers usually think about these problems these are really not separate crises at all.[1]

The UN paper contains some sobering predictions. The team argues that 'today's "dominant economic theories" and conceptions of modern capitalism are inadequate because they falsely assume societies will have continued access to cheap energy, like fossil fuels'; furthermore, these theories 'generally don't factor in sink costs – meaning costs that can't be recovered – like climate change, and they fail to account for the potential socio-political consequences that could result from continued unchecked consumption and growth.'[2] They also add that no widely applicable economic models have been developed specifically for the

---

1 Nafeez Ahmed, 'This is how UN scientists are preparing for the end of capitalism,' www.independent.co.uk/news/long_reads/capitalism-un-scientists-preparing-end-fossil-fuels-warning-demise-a8523856.html (accessed 09 August 2019).
2 https://bigthink.com/stephen-johnson/scientists-to-un-to-stop-climate-change-modern-capitalism-needs-to-die (accessed 09 August 2019).

upcoming era, which is startling given the existence of several major publications on regenerative economics, the most significant of which, to my mind, is the economist Kate Raworth's *Doughnut Economics*. So let's come back to Raworth's clear and grounded regenerative model. Raworth (2017: 4) proposes that a revolution in economic theory is necessary and begins by saying it is clear that linear, growth-economy cannot last and that what must replace it is a circular regenerative model (34). Currently, she explains (74) we use the planet as both source and sink:

SOURCE	and	SINK
whereby we		whereby we
extract materials		dump waste
(eg fossil fuels)		

A prime film example of this practice of sinking waste can be found in Soderbergh's film *Erin Brockovich*, 2000, which exposes the contamination of drinking water through a giant corporation's chemical dumping.[3] Incidentally, this practice of source and sink finds easy parallels in finance contexts in the form of PUMP and DUMP scams (as we see in the self-centred ghastliness of Gordon Gekko in Oliver Stone's *Wall Street*, 1987; the corporate greed and fraud as exposed in Martin Scorsese's *The Wolf of Wall Street*, 2013 – both films based in real stock-broker manipulators[4]). In Section 3 of this chapter, I'll be addressing these three films through the Doughnut optic.

This source and sink attitude is one by which we treat the planet as if it were an open system (like the growth model of economics) which of course it isn't. It's a closed system, a living system that thrives by

---

3 Not a lot learnt from this bad practice it would seem! Incidents of leakages of this chemical, hexavalent chromium (known also as chromium 6), into the atmosphere and into water beds have continued both in the USA and other countries (Australia, Bangladesh, Greece, Iraq). Other examples of water pollution are on-going. In 2014, in a cost-saving exercise, the city of Flint, in Michigan State, USA, switched its water authority supplier. Due to insufficient water treatment, lead leached from ageing water pies into the drinking water. In Washington DC, lead is still present at harmful levels in the drinking water, despite the fact that the danger was first alerted in 2001, and that several court cases have been brought with scientific support all of which to date have been dismissed by judges as providing insufficient evidence.

4 There is no shortage of real material for these scamming practices: *Rogue Trader* about Nick Leeson (1999), the documentary *Inside Job* (2010) and *Margin Call* (2011), both about the collapse of the sub-prime market in 2007/8.

constantly recycling that which is within (Raworth, 74 and 212); the planet is a system that is 'dynamic, self-organising, evolving, diverse and not uniform; it is messy, non-linear, turbulent and chaotic' – in short, in permanent revolution (Raworth, citing Donella Meadows, 141). Following on from this, it is not difficult to draw parallels between, on the one hand, the capitalist growth-economy model of the up-curving line and the powerful effect it has had on fixing the mind set (as a good thing) and, on the other, the classical film three-act arc of exposition-confrontation-resolution which still dominates mainstream narrative cinema today and which is a fixing way of mediating reality, of giving the impression that life is like that – that is to say, it is permanently resolvable, which it patently isn't. Why, given the dynamic complexity of the planet which 'spends its time in transient behaviour on its way to somewhere else' (141) should we be surprised that, 'when it comes to finance, stability breeds instability'? (146) Why not accept instability as the starting point?

But No, the world of economics (which has built itself into a science) 'shapes the world we inhabit' (10), it is the 'mother tongue of public policy, the language of public life, and the mindset that shapes society' (6). The economic world (as presently founded) believes in mathematically neat equilibria (checks and balances; market prices set where costs and utility meet, finding their natural equilibrium, as it were). And for that reason it continues to be 'taught as a linear, mechanical and predictable one' in which markets adjust instantly to shocks (137). As a result, events, such as the Lehman Brothers' collapse in 2008, are interpreted by mainstream economists (and in news reports) as 'exogenous shocks' – as having a sudden and external cause. Whereas, in fact, they are slow in the making and should be understood as 'arising from endogenous change' – having an internal cause or origin which slowly produces an 'accumulated pressure on the system' (141). This sounds very much like the slow invisible violence Demos speaks of in relation to the effects on the environment of extractive practices by petro- and agro-businesses (see Chapter 2).

This current macro-economic mentality produces a boom and bust culture from which we have yet to evolve. Thus,

> during good economic times, banks, firms and borrowers all gain in confidence and start to take on greater risks, which pushes up the price of housing and other assets. This asset price rise, in turn, reinforces borrowers' and lenders' confidence along with their expectations that asset values will keep on rising.
>
> (146)

However, as economist Hyman Minsky points out the 'tendency to transform doing well into speculative boom is the basic instability in a capitalist economy' (cited in Raworth, 146). Next stop, bust. Prices can't keep pace with expectations, the economy crashes, insolvency occurs – surely proving that growth capitalism is a myth, a tiger that eats its tail. And yet, what happens? A blindingly stupid belief, post-crash, in this self-same system occurs. 'Confidence gradually rebuilds and the process begins all over again' in what is ultimately 'a rolling cycle of dynamic disequilibrium' (146) – the complete opposite of what growth economics is supposed to deliver (a strong and stable system!).

Raworth argues for the need to abandon our current degenerative industrial design of today's economic model which she describes in Figure 3.1.

This open source economy model (take→make→use→lose) breaks the natural cycle of the living world. Instead of allowing the planet to continually recycle life's building blocks, we extract energy and exploit materials to make products which we then use and lose to waste: energy in the form of waste-heat and materials into waste-matter (212). We even deliberately create further waste by building-in obsolescence into our machines, thereby obliging the consumer to consume more. We also have entered the manufacturing nightmare of producing increasingly cheaper things (clothing, toys, utensils, food, etc.) – presumably in part to target the poor and combat their

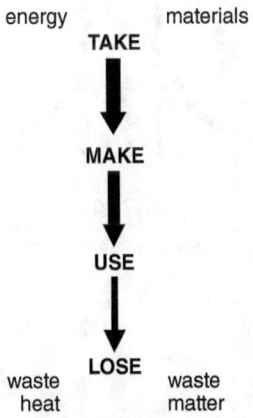

The caterpillar economy of degenerative industrial design.

Figure 3.1 Raworth's diagram of the degenerative linear economy (2017: 212).

50 *Where we are now and where we need to be*

un-buying – which has led to an increased overload of objects and thereby the throwaway culture (223).

It is time, says Raworth, to re-examine this economic growth pattern at all costs and create an economy that is distributive and regenerative by design (237). So, for example, rather than the West selling off its waste, it could employ people within its own domain to recycle it on a full-scale (including in this context plastics or nuclear waste – the two most Ecocidal waste products currently in use[5]). Currently, China is refusing to take the West's plastic waste, so we do now have to do something about it. Sri Lanka is presently refusing to process the UK's bio-medical waste because, in a recent case, it illegally contained human body-parts (are there no depths to which we can sink?).[6] Indeed, with regard to both biological and technological materials, Raworth (referring to the Ellen MacArthur Foundation's model) proposes a cyclical economy of TAKE=>MAKE=>CONSUME/USE=>RE-USE (Figure 3.2):

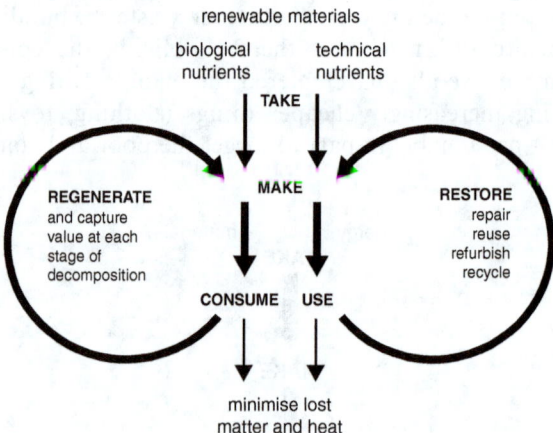

The butterfly economy: regenerative by design.

Figure 3.2 Raworth's diagram of Ellen MacArthur Foundation's butterfly model for a circular economy (2017: 220).

---

5 Intriguingly nuclear waste can be recycled, high level nuclear waste can be reprocessed to extract nuclear fuel.
6 This was reported on French radio news *France-Inter, le journal de 7h30*, 29 July 2019.

In this model, says Raworth, we consider all materials

> as belonging to one of two nutrient cycles: *biological* nutrients such as soil, plants and animals, and *technical* nutrients such as plastics, synthetics and metals. The two cycles become the butterfly's two wings, in which the materials are never 'used up' and thrown away but are used again and again and again through cycles of reuse and renewal.
>
> (221)

Just to cite a few examples of this re-use practice: in Brazil, after plastic waste contributed to deadly floods in Recife, the neighbourhood took action in a scheme that is being imitated around the world. Not only were clothes created out of the plastic bags and cups clogging up the river Tejipió, the waste removed from this same river was recycled into a house built on the river bank and called Casa Lixo (House of Trash).[7] At the other end of the designer scale, Philippe Starck warns against the over-use of raw materials and energy (they must be used as little as possible and be recyclable); thus, he prefers to use synthetic material over natural ones for his own work, reasoning that he would 'rather work with someone who uses fully traceable plastics to someone who kills trees.'[8] In yet another context, plastic composite roads (processed from post-consumer plastics and an asphalt mix) have been implemented in Australia, India, and Indonesia.[9]

The obvious starting point to a new economic order begins with recognising its dynamic complexity. Raworth points out that 'complexity thinking (is) essential to our understanding much of the world'; however, economics has been slow to recognise, let alone accept this (137). Encouragingly, there are a number of economists, including Raworth herself, who are thinking in new systems, making 'complexity economics, network theory, and evolutionary economics among the most dynamic fields of economic research' (137). Undoubtedly, for the lay person, her Doughnut Theory is presently one of the most accessible in that it offers a 'simple visualisation of the dual conditions – social and ecological – that underpin collective human well-being' (295). Indeed, her economic theory is one that is based in humanity's long-term

---

7 See 'Plastic into profit' by Sandra Laville (*The Guardian*, 12 April 2018, 39).
8 www.unenvironment.org/news-and-stories/story/capitalism-not-suitable-future-says-inventor-and-designer-philippe-starck (accessed 09 September 2019).
9 For a heartening read on examples of circular economy take a look at the special issue of *Resurgence and Ecologist* (*Regeneration*, issue No. 313, March/April, 2019).

goals – which, incidentally, go back to the original meaning of economics, as Aristotle first pointed out: 'the noble art of managing the household' as opposed to 'the pernicious art of accumulating wealth' (273).

Art – the act of making – is part and parcel of this noble art of household-planetary management. We see examples of it everywhere. The work of Ai Weiwei is indivisible from his politics, in particular his campaign for human rights, transparency, and justice in his homeland, China. Similarly, in the music of composer Outi Tarkiainen, which speaks out on behalf of the Sámi people of the northern parts of Finland, Norway, and Sweden whose way of life has been consistently undermined by successive regimes wishing to access mineral and other resources.[10] Or take a look at the paintings of Frank Bowling whose work never ceases to address disquieting political events, ranging from the effects of slavery to the violent interference of European and American governments in the process of decolonisation. Elsewhere, Kara Walker's silhouette installations and sculptures (such as *Fons Americanus*) have the uncanny ability to make her audiences see the unspeakable reality of slavery and racism.[11] Or witness Akram Khan's re-imagining of the classical ballet, *Giselle*, 'through the lens of globalisation and its imbalances of wealth, power and labour.'[12] See, in the installation works of Olafur Eliasson how 'art and culture can give language to things that are hard to articulate (such as the climate emergency), how data is often rooted in fear, whereas art is positive – it can inspire us.'[13] As a form of art and culture, cinema is no exception. As one ecosystem among a myriad of others, so too most assuredly it has its place in the practice of noble planetary management.

To this end, let us take a quick glance at the economics of this cultural industry. By adapting Raworth's developed model on National Doughnuts, in which she maps the position of some 150 countries according to their relation to Humanity's Sweet-Spot (that is, living *in* the Doughnut), we can go on to map the positioning of the major film industry countries in terms of both production and finance. To start with here (in Figure 3.3) is Raworth's mapping:

10 Listen to her *Midnight Sun Variations* (2019); *Saivo* (2017); her song-cycle *The Earth, Spring's Daughter* (2016); *Into the Woodland Silence* (2013).
11 For an excellent study of Kara Walker's work see, Gwendolyn Dubois Shaw, *Seeing the Unspeakable: The Art of Kara Walker*, Durham, NC and London, Duke University Press, 2004.
12 See Ruth Little's programme notes 'The violence of inequality: on re-imagining Giselle,' *Akram Khan's Giselle*, London, Sadler's Wells, 21 September 2019.
13 Olafur Eliasson cited in 'The bigger picture' (*The Guardian Weekend*, 29 June 2019, 24) (full page reference 24–5).

Where we are now and where we need to be 53

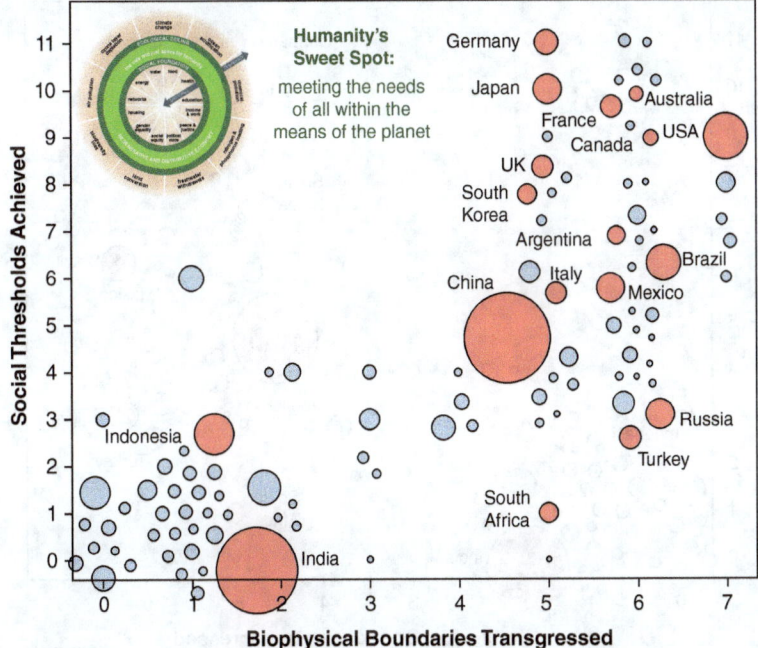

Figure 3.3 Raworth's chart of 150 nations' relation to the Doughnut.[14]

This overview reveals that, in relation to the Doughnut (humanity's place of well-being), *all* countries are developing countries 'because no country in the world can say that it is even close to meeting the needs of all its people within the means of the planet.' The only nation even close to the Sweet-Spot is Vietnam. Raworth then further clarifies this mapping by establishing three broad groupings of countries, as follows (Figure 3.4).

She draws the following analysis. Group A places countries that, whilst they are not crossing the planetary boundaries to the extent that Groups B and C are, nonetheless, fall 'very far short on meeting people's need, including G20 members India and Indonesia.' Group B clusters together middle-income, emerging economies, including the BRIC countries and Argentina (all G20 countries). These nations are 'falling short on social needs while already crossing biophysical

---

14 All quotes and diagrams were taken from www.kateraworth.com/2018/12/01/doing-the-doughnut-at-the-g20/ (accessed 04 December 2018).

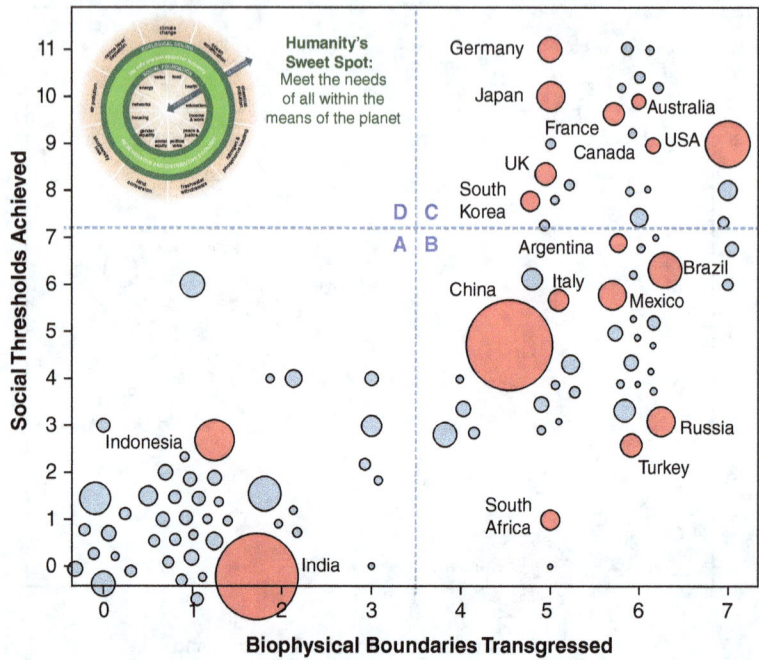

Figure 3.4 Raworth's chart of nations' relation to the two planetary boundaries (social and ecological/biospherical).

boundaries.' Group C contains 'today's high-income countries – including G20 members like the USA, the UK, France, Germany and the EU 28 itself,' nations that 'cannot be called developed, given that their resource consumption is greatly overshooting Earth's boundaries and, in the process, undermining prospects for all other countries.' And, finally, there are no countries in the 'Sweet-Spot cluster D (for Doughnut).' Currently the problem resides in the fact that all three groups seem to be moving in one direction, away from the Doughnut. As Raworth concludes:

> As the world major economies, the G20 should be leading this transformation (getting into the Doughnut), with countries starting in all three clusters. But since a key current criterion of G20 membership is having a large GPD, each country is geopolitically locked into pursuing GDP growth to keep its place in the annual G20 Family Photo.

Where we are now and where we need to be 55

But, she adds there is some hope, 'for leadership on the Doughnut Challenge, look, instead, to the Wellbeing Economy Governments (...) among them New Zealand, Scotland and Iceland.'[15]

Below, I have endeavoured to re-iterate the above groupings of economies in relation to what is currently happening in the film industry as far as the major film nations are concerned (see Figure 3.5). This is not an exclusive mapping but it does offer a preliminary method of analysis whereby we can view this particular ecosystem in relation to the Biosphere (Humanity's Sweet-Spot). My starting point was to consider film production, first, in terms of the major film nations' average film

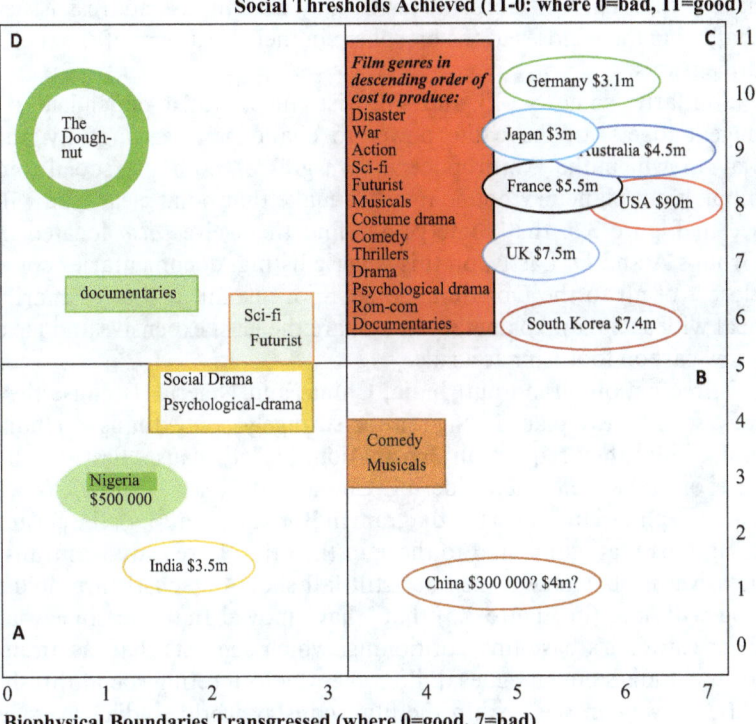

Figure 3.5 Raworth model adapted to film industry analysis.

15 All quotes and diagrams taken from www.kateraworth.com/2018/12/01/doing-the-doughnut-at-the-g20/ (accessed 04 December 2018).

budget (for example, US: $90m[16]), and, second, in terms of the generic types produced which I have ranked in order of their relative budgets. Thus, disaster movies (and indeed blockbusters) tend to be the most expensive to make. Least expensive in terms of narrative cinema tend to be dramas. This list is in the red box placed in Group C. The reasoning for this placement is that all generic forms have a carbon footprint, some more harmful to the ecological boundaries than others. Some genres in narrative terms could also be deemed to be harmful to the social boundaries insofar as they advocate violence, destruction or have narrative arcs that endorse the capitalist growth model to the detriment of the social thresholds. But others, such as comedy, musicals, social and psychological dramas, whilst they might (or might not) have a high carbon footprint tariff, may nonetheless address issues that bring them closer to the biosphere in their treatment of the social dimensions.

Similarly, some Sci-Fi and Futurist films, whilst dependent on huge budgets, do have the power to challenge, meaningfully, the way in which the Anthropocene transgresses both the social and biophysical-planetary boundaries. To make that point clear, you will see in Figure 3.5, that some genres find themselves *also* located in Groups A and D. Lastly, on this generic listing, documentaries come closest of all to the Doughnut Sweet-Spot insofar as they primarily deal with social and planetary issues, are the least expensive and have a low carbon footprint as a rule.

Three nations stand out: India, China, and Nigeria. Because they may seem oddly placed (Nigeria) or strangely contextualised (India and China) they require an explanation. So let's pause, first, on the case of India before considering China and Nigeria. India, as we know, is placed in Raworth's diagram in Box A (Figure 3.4), the cluster of nations that, compared to those in the other boxes, are less transgressive in planetary terms, but still fall short on social thresholds. You will note (in Figure 3.5) that I have moved India up somewhat from Raworth's base-line positioning, which suggests that, as an industry, India's cinema goes quite some way to meeting social thresholds. As we can see, within the film industry world, India's average film budget ($3.5m[17]) places it at the lower end of the other producing countries' budgets (thus apart from its big Bollywood films, all other

---

16 See https://parlaystudios.com/blog/feature-film-budget-breakdown/ (accessed 15 August 2019).
17 http://filmmakersfans.com/real-budget-range-indian-films-check-now/ (accessed 26 August 2019).

films are produced on a much smaller budget). In consumption terms, India's cinema is a success story on two counts. First, with a global box-office revenue of $2.9bn, India's film industry comes fourth in the world. Second, and even more significantly, 85% of this revenue comes from its national production ($2.1bn). It seems, therefore, that we can safely state that India's cinema seems to meet the needs of its audiences – and in this context, this aspect of India's economy is principally regenerative.

China's cinema industry is not at all easy to quantify. Precise figures are very difficult to trace. However, as far as I have been able to determine, there is the mainland average budget which, by all the accounts I have found, appears to be fairly low, at $300,000.[18] But then, there is another budget figure of $4m which this time includes China's co-productions with Hong-Kong, and possibly Hong-Kong productions themselves.[19] This lack of clarity notwithstanding, what can be determined is that the dominant genre on mainland China is drama (yearly around 179), three times the number of the next most produced types: thrillers (69), documentaries (57), action (57), and romance (48). As we can see, only one expensive genre (action films) ranks amongst the leaders.[20] Figures for 2018 indicate that the mainland industry produced 50% of the films viewed, garnering a box-office revenue of $4.5bn, which represents half of its total box-office revenue ($9bn).[21] Figures for 2019, however, show a significant drop in attendance for home products to 30% with a 70% share of the revenue raised primarily on US imports.[22] This poor showing is a result, doubtless, of the tighter censorship imposed in 2016 on film production and the tax clampdown on the industry.[23] In 2016, China passed a law banning content deemed to be harmful to the 'dignity, honor and interests' of the People's Republic and

---

18 http://factsanddetails.com/china/cat7/sub42/item245.html (accessed 26 August 19).
19 www.screendaily.com/news/chinese-box-office-down-27-in-first-half-of-2019-as-market-woes-start-to-bite/5141001.article (accessed 16 August 2019).
20 https://en.wikipedia.org/wiki/Category:Chinese_films_by_genre (accessed 16 August 2019).
21 https://variety.com/2019/film/asia/shanghai-chinese-film-industry-statistics-needs-quality-growth-1203249656/ (accessed 26 August 2019).
22 www.outsidethebeltway.com/china-to-surpass-u-s-in-box-office-in-2020/ (accessed 26 August 2019). This same article predicts that, by 2020, China will outstrip the USA in terms of box-office revenue. It estimates a return of $14.9bn, which includes a projected $4.5bn revenue on national products. However, this represents less than a third of the overall films consumed – so there is no change in current trends.
23 www.screendaily.com/news/chinese-box-office-down-27-in-first-half-of-2019-as-market-woes-start-to-bite/5141001.article (accessed 16 August 2019).

encouraged, instead, the promotion of 'socialist core values,' with all film projects to be approved by the National People's Congress Standing Committee.[24] Not exactly a flourishing environment, perhaps, for the meeting of social thresholds, to say nothing of the diminished revenue on their national product (down to $2.9bn). And, as we note, its downward trend is the complete opposite of India's. China's global box-office revenue may well place the industry in second place after the USA. However, the present drop in attendance for national productions is a timely reminder that its own industry is not exactly booming.

Nigeria's film industry is a stand-out case to consider. First, because it is most often over-looked in any reports on the global box-office revenue rating of a nation's film production and consumption. In fact, Nigeria currently stands third, as I shall explain below. So clearly, this oversight needs rectifying. Second, whilst Nigeria is the world's 20th largest economy, it is nonetheless beset by serious poverty amongst most of its inhabitants, due in part to an explosive population growth and also to poor government management of the state's finances. It is also a country that has the worst deforestation record, leading to soil degradation. These factors, added to other environmental issues (such as oil spills and the impact of drilling for natural resources), make Nigeria environmentally in considerable shortfall of both social and planetary boundaries. However, over the last 20 years, Nigeria's film industry has considerably bucked this negative trend and has become one of the most viable commercial concerns with very positive returns in terms of a regenerative economy.

Nigeria's film industry, as you can see in the above mapping (Figure 3.5), is closer, as a producing nation, to the Doughnut than any other of the ten nations listed. For a start, Nigeria's film production runs mainly to the less expensive products of thrillers, dramas, and romantic comedies. Second, better investment practices since the early 2000s (including wise state funding) have allowed for good production values. The industry produces 75% of films viewed, creating a revenue for national production of $3.8bn (and a global revenue of $5.1bn)[25] – a complete reversal of China's output and revenue, we note (as indeed was the case for India). The subject-matter is often contemporaneous, at times political even, thereby attracting good audience figures at home, across the African continent and the diaspora. A second reason, to which I shall now turn, has to do with its budget and revenue practices as opposed to those of the other nations cited.

---

24 https://en.wikipedia.org/wiki/Cinema_of_China (accessed 17 August 2019).
25 https://en.wikipedia.org/wiki/Cinema_of_Nigeria (accessed 15 August 2019).

Taking into consideration the average budgets for the ten film-producing nations, we can see that the USA is by far the leader with an average budget ($90m) 12 times that of its nearest competitor (the UK, at $7.5m).[26] Nigeria may have appeared an odd inclusion, therefore, but not so. Its average budget is indeed very low by comparison with all the other countries, mainland China excepted. However, in global revenue terms and specifically in relation to its *national film production*, it is the second most valuable film industry.[27] Globally, the USA is first, generating $43.4bn per annum.[28] Nigeria's cinema, at $3.8bn, is second. And, as we saw above, China is third with $2.9bn.[29] Running close behind is India at $2.1bn. Financially and generically, Nigeria stands as an example of an apparently good economic model. Its industrial practice provides us with a model *certainly* worth investigating because it goes towards answering Raworth's call 'to create economies that promote human prosperity in a flourishing web of life' (287). Whilst making products meeting audiences' needs (to the tune of 75%), it does so at reasonable cost, implicitly producing a lower carbon-effect than the more expensive cinema industries. Furthermore, its current industrial practices are having major social benefits. Here is what Nigerian actor and producer Charles Awurum has to say:

---

26 https://parlaystudios.com/blog/feature-film-budget-breakdown/ (accessed 15 August 2019).
27 https://en.wikipedia.org/wiki/Cinema_of_Nigeria (accessed 15 August 2019). The authors, writing in 2016, place Nigeria third (after India), but over the past three years Nigeria has surpassed India. Understanding the publication of global and national budgets within the film industry is a very hairy process, because one is never quite sure what quantitative data are being included. Thus, elsewhere in your searches you may find that another list is forthcoming with the USA/Canada counted together and coming first (with $11.1bn), China second ($6.6bn, although actually it is now currently at $4.5bn), UK third ($6.5bn), Japan ($2bn) then India ($1.9bn), and South Korea ($1.4bn); and no sign of Nigeria! These particular figures are generated by World Atlas at www.worldatlas.com/articles/largest-film-industries-in-the-world.html (accessed 15 August 2019). In this particular instance, however, the figures appear to refer, first, to the internal box-office revenue and, second, not always to the specific national product revenue (the USA excepted). So in terms of the UK figures the $6.5bn includes revenues from US films (only 37% of the films are British), the same for China (currently only 30% of the films are Chinese, down from 58% in previous years), Japan (54%), and South Korea (52%). In terms of checking the global revenue of each industry-based nation's production, the sources that I have used have provided me with the figures I have quoted in the main text.
28 https://deadline.com/2018/07/film-industry-revenue-2017-ibisworld-report-gloomy-box-office-1202425692/ (accessed 15 August 2019).
29 https://en.wikipedia.org/wiki/Cinema_of_India (accessed 15 August 2019).

the impact of Nollywood on Nigeria is there for all to see. If for nothing else, Nollywood has created thousands of jobs for so many Nigerians. The industry is open to all who are talented in all areas of the motion picture industry. It has drastically prevented and reduced the crime rate in the country, put food on people's table – and the multiplier effect is tremendous.[30]

Nigeria's film industry contributes to the well-being of the social sphere by generating jobs (particularly for the young) and contributing to its country's GDP; it has established cinema theatres all over the country allowing less wealthy areas to enjoy film, making it an affordable product; it offers narratives that speak to the needs of the social threshold – at least in cultural terms, and through generic types of films that for the most part provide socially relevant and consciously cosmopolitan narratives (see, for example, *The Figurine*, Kunle Afolayan, a 2009 thriller; *Ijé*, Chineze Anyaene, a 2010 drama about justice; *Half of a Yellow Sun*, Biyi Bandele, a 2013 faction drama set during the Nigerian Civil War; *The Wedding Party*, Kemi Adetiba and Niyi Akinmolayan, a 2016 rom-com).

The USA is of course the behemoth of the world's film industries; its global dominance is crushing in relation to all the other producing countries located in Figure 3.5. However, India and Nigeria are, in their own right, examples of a near-perfect circular economy where their film products are concerned. All the other producing countries, at best, raise only around 50% of their box-office revenue on their national products – thus losing out on domestic revenue they could raise and recycle. And, of course, monies that could have been ploughed back into the industry (as in Nigeria's model, for example) have slipped away as profit revenue to the USA.

3.2 Take→make→use→re-use

Let us now pick up once more on Raworth's advocacy of circular economy and consider it in relation to film practice. As we saw in Chapter 1, even if digital cinema does come with carbon footprint issues, which obviously need to be addressed, nonetheless, it is clear that the new era of digital filmmaking makes this cyclical economic practice possible. First,

---

30 In: Steve Omanufeme (2016) 'Runaway Success' a truly heartening article published by the IMF on Nigeria's film industry www.imf.org/external/pubs/ft/fandd/2016/06/omanufeme.htm (accessed 15 August 2019).

as a product, there is no longer the need to produce plastic/petrol-based film, nor indeed to process it chemically; furthermore, the digital space can be deleted and re-used. Thus, as a virtual-visual commodity there is, technically speaking, no waste at all. Second, as a practice, digital filmmaking comes very close to being a closed-source recycling economy model in that filming, editing, sound-tracking, special effects can all be done on a camera-computer interface. Third, as a distributive and exhibition source, digital prints cost a fraction of celluloid prints (0.025%) and there is no deterioration in a digital film's screening lifetime; it is also worth making the point that in exhibition terms these films can be digitally sourced into theatres, streamed via satellite, the internet, and, thus, have the huge potential of building knowledge (instead of privatising and enclosing it); YouTube, Vimeo, and other such platforms represent an enormous potential for filmmakers to present their work, and for individuals to share their skills (be it in carpentry, weaving or whatever); and of course the value of online documentaries cannot be underestimated in this context (on a personal note, a former Film and Practice student of mine, James Nikitine, who has been preoccupied by marine life for most of his life, has recently established a marine biology website www.manaiaproductions.com producing documentaries with an ethos of communicating the balance between nature-conservation, science, and livelihoods). Digital cinema becomes, then, part of what we could call the weightless no-waste economy (even if, as we pointed out in Chapter 1, upgrading systems and preservation does entail costs).

In terms of film practice there is a need to develop and deepen standard narratives by moving away from the three-act structure of exposition-confrontation-resolution (most of which endorse the Anthropocene/Capitalocene model) *and* to move away also from a simplified form of the three «Rs» (Return, Retribution, and Redemption). Clearly, we know of a good number of films which already have the interrelatedness of the two economic spheres (ecological and social) as a starting point – these are models to study (Ken Loach's films and those of and Agnès Varda come to my mind). Documentaries are well ahead in this domain of exposition, and represent an exciting new terrain of film practice. For example, a recent survey showed that documentaries on climate change outstrip fiction films on the same topic by a ratio of 7:2. The documentary *Demain/Tomorrow* (Cyril Dion and Mélanie Laurent, 2015) is just one recent example in which the failures of the growth model of economics are exposed by a world-wide exploration of regenerative models of economics – all of which is very heartening (as the follow-up documentary to this film, *Après Demain*, Cyril Dion and Laure Noualhat, 2018, makes clear).

62 *Where we are now and where we need to be*

A second point, in terms of film practice, is that the digital has certainly democratised access to film technology. Furthermore, given its digital nature, the disempowered *can* speak back, as is evidenced by the smuggling out, via satellite phones, of films about the Arab Spring of 2010/11; the screening, at festivals and on TV channels outside of Iran, of banned filmmaker Jafah Paneer's *Tehran Taxi*, 2015; and the ongoing practice, particularly in India, of Video Volunteers which promotes community media. To a degree, we have democratised distribution and exhibition, primarily by using the internet. We have even been fairly successful in democratising the financing of production through Crowd-Sourcing On-line (see, for example, Franny Armstrong's *The Age of Stupid*, 2009, a wonderful documentary on climate change; as is *Demain*, that was also partly crowd-funded). In all these modalities, we move away from the Capitalocene and towards the Doughnut's Sweet-Spot. In this context, moreover, documentaries are true leaders on both socio-political and environmental issues – and, perhaps coming as something of a surprise, the current major streaming entertainment company, Netflix, has seen the value of this film form, recently financing the Oscar-winning short documentary *White Helmets* (Orlando von Einsiedel, 2016) about volunteer rescue workers of the Syrian Civil Defence.[31]

As for fiction films, we need a broad range of products which, on the one hand, expose practices of Ecocide (in the manner, say, of *Erin Brockovich*), films that show the effects on our social foundation of systematised racism, sexism, homophobia, chauvinism, and other forms of exclusionism – much as *3 Billboards outside Ebbing, Missouri* (Martin McDonagh, 2017) does and, to a degree, *Hidden Figures* (see Chapter 1); and, of course, on the other hand, we also need films with heart-warming narratives of collectivism, compassion, and curiosity (the UK/French *Paddington Bear* films come to mind...[32]).

Using Raworth's model, how might this be achieved? We note, for example, that both Loach and Varda have as their starting narrative point the interrelatedness between the environment, the individual,

---

31 James Le Mesurier, who was the co-founder, with Farouq Habib, of the White Helmets Syrian Civil Defence group was found dead in Istanbul on 09 November 2019. He had apparently fallen from his apartment's balcony.
32 *Paddington* (Paul King, 2014) was produced by Heyday Films, StudioCanal UK, and TF1 Films Production; this first film's budget was $65m, and its box-office revenue was $268m. The second film, *Paddington 2* (Paul King, 2017), again produced by Heyday Films and StudioCanal UK, was made on a budget of $40m and box-officed $227.3m.

and society, and the breaches between the social and planetary ecological spheres. At times, their narrative landscape stretches to include the impact of environmental transgressions upon both society and the individual (see, for example, Varda's *Sans toit ni loi/Vagabond*, 1984; or indeed her documentary which directly engages with the regenerative uses of 'waste,' be it food or plastic as in *Les Glaneurs et la glaneuse/The Gleaners and I*, 2000). Ken Loach's *The Wind that Shakes the Barley* (2006) looks closely at the effects of the 1912–21 Irish War of Independence (itself an outcome of British colonisation in the 16th century) on a country at war, the factions between and within the two opposing sides, the impact of all this upon the social infrastructures, and on the individuals involved. These are rich tapestries to examine as models.

And as with tapestries we must first understand the threads that compose them. As Raworth says, what is key in all of this is who controls land, money-creation, enterprise, technology, and knowledge (177). In Chapter 2 we examined two kinds of wealth creation: land ownership and exploitation (*Tulsa* and *Giant*) and the impetus to always accumulate more (*Mildred Pierce* and *Tulsa*). Presently, a new brand of land ownership (governments selling off land to private developers) has evolved as well as new forms of money-creation (known as the GIG economy). Both are forms of weightless capitalism. Both practise a form of levying capital without any input or effort. Land has become financialised (often through privatisation sales, often money-laundering scams it has to be said), and the owners merely take the revenue. As Will Self points out, this 'rentier economy is driven by profits (...) meanwhile the rentier class keep right on reaping where they never sowed.'[33] The GIG economy is an equally exploitative system – as exposed in Ken Loach's devastating *Sorry We Missed You* (2019). GIG businesses such as Uber and Deliveroo pay workers rates below the minimum wage, pay no social charges, workers in Amazon warehouses suffer similar terrible conditions. Meantime the invisible owners run laughing all the way to the bank... oops, perhaps not banks because now they too, at 0% interest rates, offer nothing (I wonder where these GIGs and other web giants, known in French as GAFAs, place their huge fortunes, therefore...hmmm).

---

33 See Will Self (2018: 19) in his review of Brett Christophers (2018) *The New Renclosure: The Appropriation of Public Land in Neoliberal Britain* (*The Guardian Review Section*, 08 December 2018, 18–19).

Knowledge as we discussed in Chapter 2 has become privatised by internet companies. Alarmingly (at least to my mind as an academic), universities deliver «knowledge» in the form of pre-loaded texts online which students then access without needing to go to the libraries and explore. Technology has commodified knowledge. But, it is the social management, the ownership of technology that we must object to, not the technology itself.[34] Let's just think about this in film terms: if we compare the *Star Wars* space-swashbuckler film with *The Matrix* and *Blade Runner* and *Strange Days* we get a very diverse set of narratives about technology ownership. *Star Wars* (George Lucas, 1977), arguably, provides an example of a good society brought about by giant technology («may the Force be with you», indeed); conversely, *The Matrix* (Lana and Lilly Wachowski, 1999) and *Blade Runner* (Ridley Scott, 1982) give us examples of a society controlled by wealthy owners of giant technology in which Androids are cleverer than humans; finally, *Strange Days* (Kathryn Bigelow, 1995) reduces humanity's hopes even further with micro-technology fully controlling our bodies in the form of micro-chip implants in our brains – not so far away, given we already have neural implants!

3.3 Visualising Anthropocenic greed – *Erin Brockovich* (2000), *Wall Street* (1987), and *The Wolf of Wall Street* (2013): exposing the source and sink/pump and dump mentality of giant corporations and institutions

This time I'd like us to consider the three, fairly contemporary, mainstream American movies already mentioned in this chapter, *Erin Brockovich*, *Wall Street*, and *The Wolf of Wall Street* – all three of which are based upon real people and true events. Again, I recognise that my source material is heavily weighted towards Western (indeed American) film examples, but I hope the reader understands that these choices are motivated by my expectation that the narratives are familiar to you. This, in turn (I trust), should make it easier to show how to work with the Doughnut theoretical model.

The eponymous protagonist of *Erin Brockovich* is a single mum of three who is down on her luck. The year is 1993. She is unemployed, unskilled, and without formal degrees. The film opens with a number

---

34 After all, as the economist Varoufakis (2017: 120–22) points out, the greatest exchange value today is not the technological object itself, but by far the Intellectual Property Rights – for example, iPhone.

of vignettes establishing how close she is to destitution. Her food cupboard is bare, her house is crumbling and is infested with cockroaches; to add to her misery, she suffers severe neck injuries when a car runs through a red light and crashes into her very old vehicle. In terms of the social ceiling, she is in shortfall on three if not four of the minimum social standards: education, income and work, housing, and food.

However, these vignettes also establish that she has smarts. She has knowledge of geology, maps and an excellent photographic memory. After she loses her case for damages – mostly thanks to her confrontational behaviour under cross-examination – she takes her lawyer, Mr Masry, to task and insists he give her a job. Reluctantly, he does so, making her his legal clerk. She also has, much as the real Erin Brockovich, a very sassy mouth (the real Brockovich terms it a 'potty mouth' and refers to herself as a 'trash talking crusader'). In that same vein of truth to the original persona, Erin wears clothes that do much to flatter her cleavage (or 'boobs' as she puts it) and her very long elegant legs. As the real Brockovich explains: 'yes I did dress that way. I was actually taken back by the response of many people regarding my wardrobe. I just dressed that way because it was fun and I liked it.'[35] That's what the fictional Erin retorts to Masry, word for word, when he suggests she rethink her wardrobe because the other 'girls in the office' feel uncomfortable about what she wears.

Claiming to be overwhelmed by the messy state of a real estate case-file, Masry tosses it onto Erin's desk for her to sort out. The case is brought by the Jensen family, residents of Hinkley (California) against the giant corporation Pacific Gas and Electric Company (PG&E) who want to foreclose on buying the Jensen property. As Erin leafs through the case-file, she comes across a number of medical records detailing blood test results and treatments for tumours and lymphomas (paid for by PG&E). Surprised at the presence of these records in an ordinary real estate case, Erin seeks out the Jensen family only to discover that PG&E had covered the medical costs 'because, as Donna Jensen explains, of the chromium' (which the company had claimed was safe). Erin takes it upon herself to delve deeper. She gets hold of the local water-board records and does some water testing at the PG&E site to ascertain what is going on. Testing reveals evidence of water contamination with hexavalent chromium (Cr6) and the whole scandalous cover-up starts to unravel, namely, that PG&E were using Cr6 (a

---

35 Erin Brockovich (2010) 'Erin Brockovich – Did the movie really get it right,' www.cnbc.com/id/25790156 (accessed 29 September 2019).

known carcinogenic, and therefore not safe) in a cooling tower system to fight corrosion at their Hinkley compressor station and the contaminated water used in this process was discharged into unlined holding ponds at the site which then, in turn, leached into the groundwater source.

Erin pursues her research and talks to other Hinkley residents, only to discover that they too have been affected by this so-called safe chromium and been treated medically at PG&E's expense. Legally, there is a case to answer and Erin gathers the residents' testimonies (some 634 plaintiffs in all) in preparation for a major class-action lawsuit. However, as Masry explains, the lack of evidence that the PG&E headquarters in San Francisco had any knowledge of the problem means that the corporation could stall court proceedings for years, because the case for negligence was specifically that of the Hinkley site. Only compensation by binding arbitration is possible, says Masry (meaning a much smaller pay-out). There is, however, a final twist in this convoluted tale. A former employee of PG&E, who had been ordered by the company to destroy some documents, had in fact not bothered to do so because he was 'too lazy.' He'd been unaware of their significance until Erin's investigation, but, now he understands how key they may be to the case, he hands them over to her. The documents prove that headquarters knew as far back as 1966 that the water was contaminated with Cr6. Three years after Erin began to research the case, the plaintiffs win a huge settlement ($333m, the highest ever at that time). And Erin receives a bonus payment of $2m.

The film follows the standard format of the three-act arc structure within which a series of oppositions get played out (primarily around class, gender, and justice). Indeed, in its Hollywoodness, we might argue that it brings nothing new to the table. The title of the film makes it clear that the narrative is woven around an individual who manages to save the day. Granted, the fact that the focus is on a single woman makes it a bit more unusual for Hollywood. Even so, the resolution is a familiar trope within mainstream cinema: an individual can fight injustices done to the poor, can pull herself up by her bootstraps, and gain respect and self-esteem, and money and a nice boyfriend. However, what works *against* being dismissive of this narrative is the disruptive behaviour of Erin Brockovich herself – including her language and dress code. Her incivility towards institutions of privilege marks her out as an embodiment of resistance to discriminative practices. Through her, we see how institutional representatives (judges, lawyers) perceive her forceful language as rude and offensive, and her general comportment as unfeminine and lacking in 'charm' (as Masry puts it). Nonetheless, her manner gains her access to the Hinkley plaintiffs (who embrace her as one of them, as opposed to the posh female lawyer who exudes privilege but no understanding, as a result of which her attempts to side-line Erin completely fail). Erin's dress code, which many in the legal profession deride (lawyers, judges, secretaries), is in fact a primary tool through which she can befuddle (even hypnotise) men into letting her access the information she wants ('it's called boobs,' she retorts to Masry who asks how she managed to get past the water-board official). In short, Erin uses the very things that should exclude her from the institutions of privilege to expose their classism and sexism – and she does so knowingly (as we see in her interactions with the posh corporate lawyers Masry has brought in to help with the case[36]). Furthermore, because she is curious, she seeks out answers, acquires knowledge, and becomes very knowing of her rights and freedom to information.

What emerges in terms of the Doughnut model, then, is an interesting cyclical pattern. As a woman and mother of three kids in shortfall of the social foundation, Erin is desperate to find a way for them all to survive. Her first chance to legally find justice and retribution (over the whiplash case) is dashed to pieces by the defence lawyer's misogynistic cross-examination against which she vehemently protests. How on earth can the fact she has been married twice and divorced have any bearing on the fact she was badly injured in a car crash not of her own making? But it does, because of the language she uses, and the judge rules against her. Her second chance to find economic security comes when she forces Masry into giving her a job. Once again this is within the legal institutional framework, but this time there is a positive outcome. Erin's determination to prove her worth propels her into the world of the Hinkley residents who are, themselves, in shortfall of the social foundation – this time because of the transgressions of the ecological foundation by PG&E. Their health has been assailed by chemical pollution of their water supply, and their homes are under threat from the buy-out scheme the giant corporation PG&E is seeking to impose. Erin's drive to acquire knowledge is what ultimately pieces the puzzle together – and gives her the power to challenge and outsmart the posh lawyers PG&E have employed, as well as the two smart-suited corporate law lawyers Masry had brought in. Moreover,

---

36 For example, when asked by the female corporate lawyer how she managed to get 634 signatures from the plaintiffs (something she had signally failed to obtain) Erin retorts that she gave them all blow-jobs – a joke not lost on Masry but which shocks to the two stuffed suits.

on both a social and private level, Erin gains a sense of gender equality, first, by Masry recognising her worth and showing respect; second, by her boyfriend George acknowledging her equal rights to work. As she said to him when he asked her to choose between him and her work, 'don't ask me to give it up; all I've ever done is build my life around what men want' – not this time! No longer unskilled and with the backing of her boss, Masry, Erin has asserted a just space for both her plaintiffs and herself and her family – and this, against a mighty and disturbingly powerful giant corporation and the equally formidable legal institution (corporate law) that protects its own.

This last point about institutional power shows, sadly, how pyrrhic this victory is ultimately, since PG&E have continued to act with total disregard for anything other than profits. Their disastrous state fire record over the past ten years alone stands as a terrible indictment of this company, which nonetheless is allowed to continue functioning. Most recently, the California company was found responsible for the catastrophic wildfires of 2018, and yet it is using all the legal ruses possible to avoid a huge pay-out (including filing for bankruptcy and turning to Hedge Funders to return them to solvency). And again, in November 2019, PG&E is once more implicated in the California state fires.[37]

But to return to *Erin Brockovich*, even if, in terms of the social foundation, justice, however briefly, is within the safety of the biosphere, it is also the case that the exploitative nature of the legal profession is exposed at numerous levels, making the Doughnut theorist's eyebrows twitch, to say the least. For it is not just the giant legal firms that take the money, it is also the small ones, such as Masry and Vititoe, who can claim 40% on any settlement (though, to his credit, Masry, unsure he will win, takes on this case *pro bono*). Erin, who at first queried

---

[37] www.mercurynews.com/2019/09/29/editorial-pge-bankruptcy-plan-options-awful-or-dreadful/ (accessed 01 October 2019). And for a trace on PG&E's recent appalling track record on safety issues, the St Bruno gas explosions of 2011, the numerous fires in Butte county in 2015, the 2017 Tubbs Fire and several other state fires in 2017, and the Butte Fire of November 2018, see www.mercurynews.com/2019/09/09/pge-files-bankruptcy-exit-plan-that-caps-payments-to-wildfire-victims/?li_source=LI&li_medium=right-rail-rec_v2&li_tr=right-rail-rec_v2 (accessed 01 October 2019). And, yet again now in 2019, PG&E are deeply implicated in the most recent fires of October–November 2019 in California (see: Susie Cagle, 'California burning,' *The Guardian*, 02 November 2019, 37). In other contexts, a similar story to *Erin Brockovich* is taking place in Reserve Louisiana where the pollution caused by Pontchartrain chemical works plant's production of neoprene has been linked to a series of severe diseases in the community (see: Oliver Laughland, 'Reporting on cancer town,' *The Guardian*, 16 November 2019, 57).

this high percentage, later comes to accept and even defend the 'necessity' for these rates. And as we see, at the end of the film, Masry and Vititoe have moved into a much more up-market building, thanks to winning the case; and within the narrative curve, this is seen as a good thing, a merited outcome for all the hard work. As far the legal establishment is concerned, then, the display of wealth is crucial as a sign of success. In a culture that endorses a growth-economy, we can see how problematic the workings of the Capitalocene are – and how difficult it is to change a fixed mentality: the desire to be *manifestly* in the big league of institutional power runs deep. Where this mentality (what I've referred to in Chapter 2 as deregulated madness) starts, is of course key – especially if we are to find a way to change.

This inability or refusal to «change» on the part of giant institutions is also at the heart of the two films about Wall Street stockbroker practices, *Wall Street* and *The Wolf of Wall Street*. Let us now consider what a Doughnut-model processing of these two narrative produces. I will not need to offer synopses since both stories tell basically the same tale: the rise and fall of a stockbroker who believes in the dictum that 'Greed is Good.' In the first film, *Wall Street*, Bud Fox has aspirations to become like his 'idol' Gordon Gekko whose own trajectory from junk-bond trader to multi-millionaire sets the model Bud wishes to emulate. Impressed by a first piece of insider information Bud passes onto him, Gekko takes Bud under his wing (seeing in Bud a young Gekko, doubtless) and thus sets him off on the illegal path of insider trading. Only towards the end, as Bud witnesses, close to home, the effects of Gekko's ruthless nature, does he finally react and mount a scam himself to prevent Gekko from destroying the company Blue Star Airlines his father works for. Bud is eventually arrested by FBI's Fraud Squad, and he agrees to turn state evidence on Gekko. In the event, the airline becomes an employee-owned company: the workers are the stakeholders; the one good outcome in this harsh narrative of greed – a 'democratising of ownership' that is 'distributive by design' as Raworth puts it (2017: 174 and 177).

It's the same rags to riches trajectory in *The Wolf of Wall Street*, a film which begins in the late-1980s period of *Wall Street* and continues through until 1999: Jordan Belfort starts at the bottom of the stock-trader pile, selling junk bonds. He is as merciless as Gekko before him. Recognising that he can persuade people of small means to buy penny-stocks (really junk bonds of no value, 'selling garbage to garbage man' as he puts it) he makes himself rich on his percentage takes (50% commission). He then decides to practise the same ethos on the wealthy at which point he moves into grandiose offices and proceeds to become a multi-millionaire through his «pump and dump» scams. As Bud and Gekko before him, he finally has his come-uppance with the FBI's Fraud Squad; but as with Gekko he only serves a minimum sentence and is soon on the trail of making money again (through his company Global Motivation Inc.). Similarly, Gekko once released from prison goes back to his old rogue-trading ways.

Gekko is based on a «type» of Wall Street manipulator and is a composite of a number of Wall Street brokers; Jordan Belfort is the real-life eponymous fraudster of the latter film. Both protagonists are involved in selling something that is not there – an alternate reality. 'It's not real, we don't create anything' says Belfort's money-crazed manager, Mark Hanna. The game is to persuade people to buy stock values that, to all intents and purposes, do not exist. So, with penny-stocks, for example, the rogue-trader sweet-talks the client into buying the hype and gets rich on that hype-sale, whilst the client has bought just hype, that is just air (hype makes 'the illusion real' as Gekko says). As long as no one sells, no one knows the fraudulent stock-raider/rogue-trader has been selling nothing. And if you can sell air why not sell toxic air while you're about it! One can easily see how this selling of alternate realities, which so easily slips through the legal nets (such as they are), feeds into the American meritocratic dream of wealth even for the «little guy». Except that, precisely, it's the opposite that happens – as we witnessed in the scandalous sub-prime mortgages racket, in which thousands of low-income home-owners lost everything.

A similar selling of air occurs with the illegal practice of insider trading. Insider trading involves profiteering on stocks, often in a failing company or one facing a takeover bid (IPO). The procedure is for the stock-trader in the know to buy the shares at a low value, but then sell-on once the company is engaged in an IPO bid and stock value rises. The rogue-traders pocket the sales commission. Other practices (as we witness in both films) are for the scamsters to buy low, sell high (taking commission), then leak information so the stock value falls when it is virtually too late for the purchasers to sell; meantime, a shell company, owned by the original fraudsters, buys up the depleted stock value and then trades it onto a willing buyer of the nearly destitute company (doubling their money as it were). It's all a game, a victimless crime as far as these rogues are concerned – because all they can see is the 'nobility,' as Bud Fox puts it, in being rich. His actual words, 'there's no nobility in poverty,' are reprised word for word by Belfort in *The Wolf of Wall Street*, showing what an indelible «truth» this is in the stockbroker mindset. 'Greed is Good' echoes Gekko. It's all a bit of a carnival, 'keep the clients on the Ferris Wheel' says Belfort's manager, adding 'money makes you a better person.' The function of these stock-raiding air-selling rogue-traders is to keep this balloon or bubble of fakery inflated, keep the hype going and their wealth accumulating (preferably in off-shore accounts). *They* know, only too well that once the bubble bursts (or is pricked by a loss in confidence) nothing is left. As *we* know only too well, there is no victimless crime (the impact of the investment bank, Lehman's crash of 2008 being but one example). The knock-on effects for all except the rogue-traders are dire: people lose their jobs, tax-payers are obliged to redress the damage done by fraudsters thus throwing countries into recession – and poverty escalates. At least 3 of the 12 dimensions of the *Social Foundation* on Raworth's Doughnut model are transgressed: income and work and social equity. And, yet, still nothing changes. Fraudsters serve a minimum sentence and unscrupulously start up again (sometimes under the guise of philanthropic practice, which often masks the reality that the money invested is a result of rogue trading[38]).

The world of *The Wolf of Wall Street* is a drugs-infused world of sex, contempt towards the poor, cruelty to dwarves (as in the opening dwarf-tossing scene), demeaning attitudes to women, where 'everyday' bigotry, racism, and homophobia prevail. Women are diminished to sexual objects, and seen as no more than trophies and profligate emblems of their men's wealth. Belfort speaks of women in terms of being like publicly traded stocks (top end being 'Blue Chips,' bottom end 'Pink Sheet hookers'), dispensable commodities, therefore. This lack of respect for others is reflected in the incessantly crude language (for example, the 'f' word appears every 20 seconds), the homophobic and anti-Semitic slurs, and overall crass behaviour of the central characters (and their acolytes). The furious pace, verbal vulgarity, and sheer noise of Scorsese's film demonstrates the madness of this world of excessive, nay, gross accumulation of wealth off the back of human fools ('people don't buy stock; it gets sold to them' declares Belfort). It is almost as if the Marquis de Sade's libertines had donned suits, walked out of Pasolini's *Salò, or the 120 Days of Sodom* (1975), and transited to Wall Street – a place, as Belfort asserts, 'for killers.' And as if to make

---

38 According to the producer of *Wall Street*, Edward R. Pressman, Gekko is partly modelled on Michael Milken, an American financier and racketeer indicted for insider trading – and now a philanthropist. https://en.wikipedia.org/wiki/Wall_Street_(1987_film) (accessed 04 October 2019). Originally sentenced to ten years for violating US securities laws, Milken's sentence was later reduced to two years. Once released, he set up the Milken Family Foundation for Research into Cancer. https://en.wikipedia.org/wiki/Michael_Milken (accessed 04 October 2019).

this absolutely clear, the dominant tone of the film is nothing short of a total assault on all of our spectator senses.

Intriguingly, some critics at the time felt that Scorsese's intended message failed to land; and, indeed, as far as a section of the cinema-going audience was concerned, this was true. Exactly as with *Wall Street* (made a quarter of a century earlier), the star performance of the central protagonists of the two films (respectively, Michael Douglas and Leonardo di Caprio) was such that the effect was to give celebrity status to fraudsters, to turn villains into boardroom icons and a source of inspiration to wannabee stockbrokers, the very people both Stone and Scorsese meant to expose.[39] According to some critics, what may have meant to be a satirical shakedown (particularly in relation to *The Wolf of Wall Street*) became an irresponsible glorification of greed.[40] Which is exactly the point, in my view (as I'll explain below). Let us remind ourselves of the following words from Gekko's famous 'Greed is Good' speech delivered to the IPO meeting at Teldar Papers: 'America has become a second-rate power. I'm not for the destruction of companies but the liberation of companies. (...) Greed will save America.' When we realise that Gekko's euphemism, 'liberation,' actually means *pump and dump/source and sink* we can see precisely how this most ruthless example of deregulatory madness can be falsely embedded in a rhetoric worthy of a US President and believed in by a willingly deluded audience that cheers Gekko on, and on. As Belfort says if you 'act as if you're wealthy (...) have unmatched confidence (...) experience (...) have tremendous success, then people will follow your advice.'[41]

Rather than seeing these films as failing to get their message across, both can be said to give full visualisation to the irresponsible glorification of greed, one might even say addiction to greed. What both films achieve is a mise-en-scène of the *normalisation* of greed, of this addiction: everyone wants in (stockbrokers, shareholders, even small

---

39 See, for example, *Variety*'s review of *Wall Street* quoted in *Wikipedia*'s article on the film: https://en.wikipedia.org/wiki/Wall_Street_(1987_film) (accessed 03 October 2019).
40 See the various reviews cited from *The Huffington Post*, *New York Film Critics Circle*, *Los Angeles Times* amongst others In: https://en.wikipedia.org/wiki/The_Wolf_of_Wall_Street_(2013_film) (accessed 04 October 2019).
41 Gekko's speech at Teldar Papers finds some unnerving echoes on Donald Trump's presidential candidacy speech then he proclaimed: 'Sadly, the American Dream is dead.' And once elected he promised to 'make America great again,' to 'put America first.'

penny-stock holders, and frankly Wall Street itself).[42] What interests us here is how quickly (in a span of ten years, from the time of *Wall Street* to that of *The Wolf of Wall Street*) that glorification both intensifies in its delusional sense of value and accelerates in the violence of self-affirmation (primarily in the form of the coked-up, foul-mouthed Belfort, but also many of his side-kicks). As Owen Gleiberman cleverly notes in March 2009 (when the Lehman scandal was in full flow, incidentally), in speaking about *Wall Street*, the film 'reveals something now which it couldn't back then: that the Gordon Gekkos of the world weren't just getting rich – they were creating an alternate reality that was going to crash down on all of us.'[43] And so it did; and so it continues, and as it does so, so too it speeds up – so fast as to be unseizable (incomprehensible). That, it seems to me, is the crucial point about these two films. They show how we know, but somehow refuse to know. In particular, *The Wolf of Wall Street* pulls on elements of the grotesque, the burlesque, and the ubuesque to present the viewer with an unambiguous image of the Capitalocene at its most vividly vile. The fact that the charisma of the lead actor-protagonists induced hopefuls into becoming stockbrokers rather makes the point as to how the practice of this exploitative behaviour perdures. Once you accept the credo that wealth accumulation makes you 'a better person,' then it is easy to see how the drive to join the world's richest 1% (who own 45% of the world's wealth) becomes a dominating imperative force.[44]

It isn't just the addictiveness of monetary success that is exposed here (almost as an illness), but also the closely knit relationship between the fiduciary and the judiciary – both supposedly institutions

---

42 There is not a little irony attached to the fact that currently Malaysia is filing charges against Goldman Sachs bosses in what has been called the 1MDB fraud scandal and which concerns the financing of *The Wolf of Wall Street*. Malaysian prosecutors allege that

> the fraud involved 1MDB officials and the former Malaysian prime minister Najib using misappropriated funds to buy art works, and luxury real estate in New York and other major cities, as well as funding the production of Hollywood films including *The Wolf of Wall Street*.
> (See, Kalyeena Makortoff, 'Malaysia files charges against Goldman Sachs bosses in 1MDB fraud,' *The Guardian*, 10 August 2019, 36)

43 Owen Gleiberman cited In: https://en.wikipedia.org/wiki/Wall_Street_(1987_film) (accessed 04 October 2019).

44 Just take a look at this online article 'Facts: Global inequality, our world's deepest pockets – "ultra high net worth individuals" – hold an outstandingly disproportionate share of global wealth,' https://inequality.org/facts/global-inequality/ (accessed 04 October 2019).

in whom we (the people) can place our trust. Every evidence is, rather, that they serve each other's interests. Gekko and Belfort repeatedly break the fiduciary confidence placed in them by their clients, in so doing they break the law. However, the judiciary sees fit to hand out very light sentences and easily circumvented compensation orders. So the carnival continues.[45] Could it be that our deregulated mentality is incapable of change, that driven as we are by our striatum (our grasping survival mechanism) we are doomed to over-consume? Or might it be the case, as a neurologist might propose, that we are capable, as a species, of a change of mentality?

So here is by now the familiar Doughnut diagram and upon it the tracings of these three films discussed (see Figure 3.6). It isn't very heartening, but it provides us with a visual model of the overall transgressions and their consequences as well as the very few instances of humanity at work in the biosphere (namely, the democratisation of ownership for the Blue Star Airlines in *Wall Street* and the seven equality outcomes in *Erin Brockovich*) (Figure 3.6).

This diagram reproduces almost exactly the same dynamics of transgressions as those seen in Chapter 2 (see Figure 2.2). The difference, however, lies in the narratives' knowingness. For, these three films show how economic, legal, and political institutions work hand in glove to organise society for their own benefit. It's a magical triangulation (striangulation, you might say). It is, after all, as Acemoglu and Robinson state, 'the political institutions that determine what economic institutions a country has' (2013: 43). It is also the case that political institutions pass the laws that govern a country and so determine the legal institutions that a country has. As a result, the way that political power is exercised is as much influenced by economic as it is by legal institutions. This dynamic makes clear how the regulatory system of a country is monopolised by a narrow elite, and how the inevitable outcome is, at best, a diminished state (at worst a lack) of equality and political rights for the majority. Here's how the loop whizzing round the triangulation works: 'political institutions enable the elites controlling political power to choose economic (and legal) institutions with few constraints or opposing forces. They also enable the elites to structure future political institutions and their evolution' (81, parenthesis mine). Trying to fight against this evidence is a bit like

---

45 A neurologist might deduce that we are, as a species, driven by our striatum, our grasping survival mechanism which pushes us to over-consume. Clearly, a change of mentality is necessary if we are to survive ourselves.

Where we are now and where we need to be 75

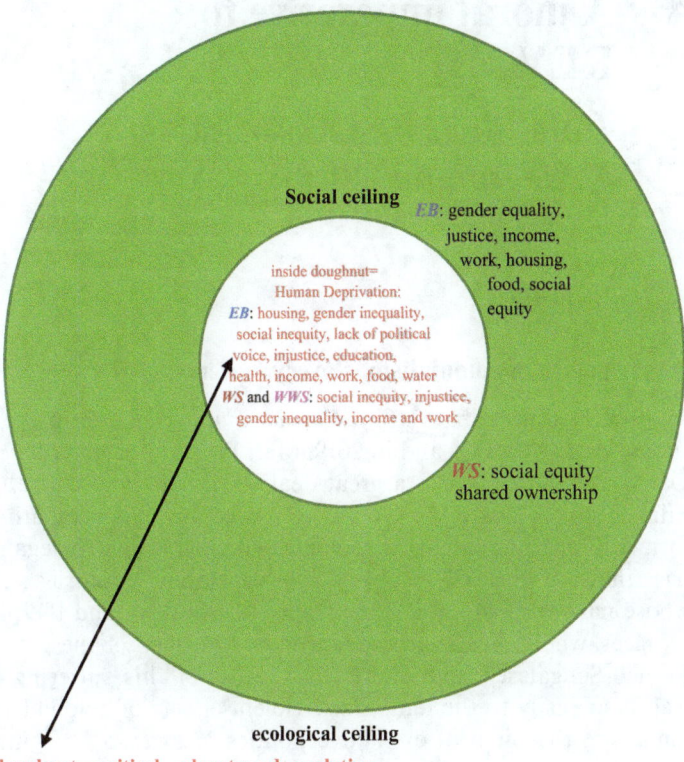

Figure 3.6 Locating *Erin Brockovich (EB)*, *Wall Street (WS)*, *Wolf of Wall Street (WWS)*.

trying to cut off Hydra's head. Every time a small victory for justice is won (as with *Erin Brockovich*) and Hydra's head is lopped off, just as quickly Hydra produces another (if not several heads) and the battle begins all over again.

In the next chapter, I propose to open these issues into broader contexts still and examine the effects of colonisation and its histories on the social and ecological boundaries of Raworth's Doughnut through two films: *Hotel Rwanda* (2004) and *Capharnaüm* (2018).

4 A moral imperative to REVOLT
Hotel Rwanda (2004) and *Caphernaüm* (2018)

4.1 Legacies of colonialism: slavery, poverty, civil wars

In 2018, Nadine Labaki's film *Caphernaüm/Chaos* (set in the Beirut slums) won at Cannes and in 2019, Mati Diop's film *Atlantique* (set in Dakar) also won. It says a great deal when two consecutive Cannes Film Festivals award the Grand Jury Prize to films emanating from former French mandates or colonies (Lebanon and Senegal, countries that are presently economically developing countries), to films whose narratives are about the tragedy of poverty and the plight of refugees, where the directors are women (the one Lebanese, the other Franco-Sengalese), and where the sources of this suffering can be traced so easily to the long, slow, violent effects of colonialism. Colonialism, the birth of extractive politics *in extenso* – beginning in the 16th century, with one of the major eventual outgrowths (in Capitolocene terms) being the Industrial Revolution which brought in its wake the birth of the age of the Anthropocene. Moreover, the triangulation I spoke of in the previous chapter between the political, economic, and legal institutions ensured that this process of colonisation was anchored within a framework of ownership and privilege. And, of course, these colonised territories had imposed upon them those very same political, economic, and legal institutions of the colonising country – a completion of a vicious circle from which those who were colonised could not escape. Small wonder then that, in *Twelve Years a Slave* (Steve McQueen, 2013), the central character, Solomon Northup, is so bewildered when he discovers that, unbeknownst to him, his status of a born-free African-American and gentleman living in the Free State of New York means nothing to the white conmen who hoodwink him into travelling to Washington DC (a Slave State), drug him, kidnap him, and sell him into slavery. And once in the Slave States he has no legal recourse – no one listens to his claims.

Instead of recognising that this rush to plunder, to take, and exploit – namely colonialism – constituted an abuse of power, the very opposite occurred. Colonialism became the passport to a nation's greatness, rather than shame. Slavery in America began as far back as the 15th and 16th centuries, with the arrival of the first colonisers from Spain. A tradition that continued with the British colonisers of the 17th century and did not end until late into the 19th century. Upon Independence, America maintained the 13 existing states as Slave States. By 1804 all Northern States had abolished slavery, but not so the Southern States. The District of Columbia did not abolish slavery until 1862, and Alexandria – a city in DC – was a major market in the American slave-trade (doubtless where Solomon Northup was traded).

Even though no legislation was ever passed in England that legalised slavery, Slavery, which began as an economic institution (slave-trade), subsequently became established as a legal institution of human chattel enslavement (a very precise example of the political-economic-legal triangulation working to the advantage of the elite).[1] British merchants traded goods for West African «slaves» captured by local rulers, they then sold the 'slaves' onto plantation owners in the American colonies and that currency was used to buy crops and commodities to bring back to Britain (the so-called Triangular Trade). Slavery became a source of huge wealth to slave owners who justified their ownership by invoking the fact that the Bible sanctioned Slavery and that it was their Christian duty to preach the scriptures to their slaves.[2] It might also interest you to know that France used the abolition of Slavery (voted for after the Revolution, but long in being abolished) to expand their colonised territory of Senegal under the pretext of spreading the abolitionist doctrine.[3]

Other examples of this political-economic-legal triangulation can be found in the British East India Company (EIC), originally formed in 1600 under Royal Charter, to trade in the Indian Ocean region. The government for its part became a joint stockholder in EIC (in 1657). By the 18th century, EIC had seized control of huge areas of the Indian subcontinent and, by the 19th century, expanded its territorial ownership by colonising parts of Southeast Asia and Hong Kong. This powerful economic institution instrumentalised its control over India thanks to a huge private army, easily shifting gears from trade to territorial ownership. And, once the British Empire expanded, the

---

1 https://en.wikipedia.org/wiki/Slavery_in_Britain (accessed 11 October 2019).
2 See https://en.wikipedia.org/wiki/12_Years_a_Slave_(film) and https://en.wikipedia.org/wiki/Slavery_in_the_colonial_United_States (accessed 11 October 2019).
3 https://en.wikipedia.org/wiki/Senegal (accessed 12 October 2019).

company effectively ruled India for a century (1757–1858), after which the British Crown assumed direct control. From the very beginning, the company was endorsed by the political and legal institutions of Britain as colonial ruler in their name.[4]

Colonialism, creations of Empires, the grabbing of territories – all are for the taking, thanks either to superior technologies of war or advanced trade practices. Whichever practice is implemented, both occur either at the expressed will or the collusion of the colonising nation's government or ruler. Thereafter, international conflicts are waged in defence of these Imperial holdings which often end up as spoils of war. A particularly horrible example of this is World War One when parts of the former Ottoman Empire and German colonies in Africa were reassigned to conquering allies. Turkey ceded Lebanon and Syria to the French and Iraq to the British. Germany ceded Rwanda to Belgium, the Cameroon to France and Britain, South West Africa (now Namibia) to Britain, and so on. We are still feeling the effects of all of this. Just to cite one instance since it concerns the film *Caphernaüm*: the hugely unsettled situation in the Middle East is not a recent phenomenon; it stretches back and through both its own history and that of its colonised and post-colonial history. Syria, Iran, Lebanon, and Iraq were former territories of Turkey's Ottoman Empire, subsequently re-colonised/mandated under the British and the French – a layering of colonial systems which impacted severely upon these countries' futures. As Acemoglu and Robinson explain:

> they were all provinces of the Ottoman Empire which heavily, and adversely, shaped the way they developed. After Ottoman rule collapsed, the Middle East was absorbed into the English and French colonial empires, which, again, stunted their possibilities. After Independence, they followed much of the former colonial world by developing hierarchical, authoritarian political regimes with few of the political and economic institutions that (…) are crucial for generating economic success. This development path was forged largely by the history of Ottoman and European rule.
>
> (2013: 61)

The tragedy is that post-colonial nations continued with the extractive policies of their former colonisers (for example, oil, gold, and

---

4 The Dutch West and East India Companies did much the same albeit on a smaller scale (for example, Dutch Cape Colony now South Africa, Indonesia, Dutch Guiana now Surinam). The French likewise.

diamond mining) and failed thereby to pursue other technologies and practices that could have enriched the countries as a whole. Again Acemoglu and Robinson paint a clear picture of how this whole process has endured:

> The Ottoman Empire was absolutist (and) the economic institutions the Ottomans imposed were highly extractive. (...) The consequence was that at the time of the Industrial Revolution the economic institutions of the Middle East were extractive. The region stagnated economically. (...) When European control ended, the same dynamics (as we later see) in sub-Saharan Africa took hold, with extractive colonial institutions taken over by independent elites. In some cases, such as the monarchy of Jordan, these elites were direct creations of the colonial powers, but this too happened frequently in Africa.
>
> (120–21, parentheses mine)

New rulers saw the advantage in continuing the colonialist process as a way of becoming wealthy at the expense of the nation and holding onto power. The result: 'poor countries are poor because those who have power make choices that create poverty' (68). Also contributing to this poverty was the effect of the deliberate strategy of 'uneven dissemination of industrial technology and manufacturing production' during the height of colonialism in the 19th century (51). The failure of the coloniser to introduce new technologies (that did not suit their extractive purposes) and impose an infrastructure for land management (that would have enabled a nation to feed itself) was, in many cases, not reversed by the decolonised nations' leaders (51–3).

Civil wars in decolonised nations, whilst emanating from a convergence of complex issues (tribal, religious, and so forth), are an outcome of this continuation of extractive economic institutions, of elitist autocratic rule that seeks only to preserve its power. The Arab Spring Revolution of 2010–12 began in Tunisia in response to oppressive regimes and poor living standards and spread across the Middle East. By far the worst autocratic response to this uprising was Syria's leader Bashar al-Assad whose brutal repression very quickly developed into the ongoing Syrian Civil War. The consequences of unleashing this war have been catastrophic on Syrians. Over half of the population are requiring humanitarian assistance: 7.6m are internally displaced, nearly 7m have sought refuge outside their country, most of whom have been placed in refugee camps (Turkey has the most with 3.6m refugees; the Lebanon has 1.5m; Jordan 1.2m; Germany three quarters of a million; the rest spread out in their thousands over some 42 countries).[5]

When we add the ruthless exploitation (just in terms of the African continent[6]) by the British, French, German, Dutch, and Belgian Empires, to the damage caused in the aftermath of decolonisation – not just in political terms (as detailed above) but also in ecological terms – it is only then that we can understand how the dreadful consequences of poverty and migration have transpired. The prospering nations of the northern hemisphere – thanks largely to the Industrial Revolution (made on the back of colonial extractive policies) – have so decimated the protective ozone layer of our planet that it is the poorest countries that are the greatest victims of its effects. Climate warming has made certain countries uninhabitable, their lands unproductive. We use up the planet's natural ability to sustain our needs by eight months into the year, meantime the most populous parts of the world are increasingly uninhabitable. Migration is the inevitable outcome, as indeed is radicalisation.

As just one example of the extreme effects of this warming let us consider the Sahel Strip (the transitional zone between the Sahara and sub-Saharan nations), also known, since the mid-1980s, as the Hunger Belt. Countries such as Mali, part of whose territories are in the Sahel Strip, are facing the consequences of climate change *in extremis* with a progressive desertification causing soil erosion and exhaustion. It is a complex picture with serious outcomes. First, the change in the rains over the past two decades, coming later and leaving earlier, has led to impoverished crops. The progressive deforestation to make way for cotton production has added to the desertification. Meantime, population growth has meant a big increase in poverty, malnutrition, and inadequate hygiene. Conflicts between agricultural communities and ethnic groups of the North and South have led to self-styled militias, which, in turn, have created fertile areas for recruitment to Islamist groups. Armed conflict in Northern Mali in 2012 between, on the one side Tuareg secessionists and radical Islamist groups and, on the other side, the government forces, has led to a large-scale internal displacement of Maliens (estimated at 300,000) and a further migration out of the country of

---

5 https://en.wikipedia.org/wiki/Refugees_of_the_Syrian_Civil_War and www.la-croix.com/Journal/refugies-syriens-monde-2018-08-22-1100963121 (accessed 14 October 2019).

6 The African continent is not alone in this of course. We are fully aware that other formerly colonised countries are suffering similar difficulties.

some 175,000 refugees.[7] Finally, on a population of 19m, humanitarian crisis affects some 7m Maliens (50% of whom are women).[8]

4.2 Legacies of colonialism: displacement, separation, non-spaces, non-beings, racisms – two visualisations, *Hotel Rwanda* and *Capharnaüm*

Before we start discussing these two films, I should explain that the methodology in this section will be shifting gears and taking a more expansive look at these films through the optic of the Doughnut. To that end, the film narratives will be located within the social and ecological/socio-political-economic contexts, as presented in the films, alongside with the layered histories that have contributed to these current contexts and to which the films may or may not refer. Earlier chapters have set the tone, this time we go deeper (see what's there and, indeed if it's the case, what's missing). This time, we will see how each film produces a series of mappings onto the Doughnut (for example: with *Hotel Rwanda*, the 1994 Genocide and the differentiated historical moments against which the Hutu/Tutsi conflicts can be read). The idea behind this is to show just how slow the violence can be in its arising and just how far the extractive effects of colonialism can reach. I hope by now that the value of using this model is becoming more evident to readers who are keen to introduce eco-theory into their film analysis. But I also have in mind that you, the reader, might see the value of using this model when it comes to film-scripting and might imagine doing so in this vein so that we don't lose sight of what Kimberlé Crenshaw calls Intersectionality – a concept closely aligned to Raworth's Doughnut theory and one which argues that the various forms of humanity are complexly interwoven, not just in terms of identity (class, race, sexual orientation, disability, and gender), but also in terms of world systems such as politics, legal, economic, and social practices and procedures – which are often embedded in *privilege* (Crenshaw, 1991).

Let us first consider *Hotel Rwanda* (Terry George, 2004). This film is set in 1994, at the peak moment of the Civil War between the Hutu and the Tutsi, during which 70% of the Tutsi were massacred by the Hutu – a violence which also manifested itself in using rape as a weapon of war, deliberately infecting women with HIV, and the

---

7 For more detail see www.iom.int/countries/mali (accessed 28 November 2019).
8 Ibid.

mass-murder of children so that there would be no future generations of Tutsi. Based on true events that occurred during the 100 days of the Rwanda Genocide, the film relates the story of Paul Rusesabagina, a Hutu, who is employed as the manager for the Belgian company Sabena's exclusive hotels in Kigali – first, The Diplomat and, shortly after the film opens, the very prestigious Hôtel des Mille Collines (so named after the soubriquet for Rwanda, land of a thousand hills). Paul lives with his wife, Tatiana (a Tutsi) and three children in the nearby comfortable suburbs. Tensions between the Hutu and Tutsi communities are running high, however, because of the extension of the UN peace-keeping mandate in Rwanda. The Hutu hardliners are opposed to any further intervention and are campaigning to take things into their own hands – as indeed we hear from the very beginning via the various broadcasts on the Hutu-Power Radio Station (Radio Télévison Libre les Mille Collines) with declarations such as the 'Tutsi have taken our land,' 'Rwanda is Hutu land,' and 'these cockroaches are traitors and invaders.' The assassination of the Rwandan President, Juvénal Habyarimana, triggers the programme of the Tutsi extermination, implemented by the military and primarily carried out by their trained militia groups, the Interahawme and the Impuzamugambi. As the genocidal campaign begins, Paul brings his family and neighbours to safety in the Hôtel des Mille Collines, where a small contingent of UN troops guarantees a level of security. Over the ensuing days, the hotel becomes a sanctuary for some 1268 Rwandans made up of Tutsi and Hutu at risk, elites as well as people of all backgrounds, including children already orphaned by the ongoing massacre. As the UN chief in command, Colonel Oliver shamefacedly admits, the UN presence is merely that of observers. Because of earlier murders of UN troops (in Somalia), they have been ordered not to fight, so they are effectively toothless. The Hutu militia exploit this, knowing that if they kill UN soldiers the peace-keeping force will be obliged to withdraw. This happens. The UN are ordered to pull out and take only the White residents of the hotel with them. Appalled at the explicit racism of this order, Colonel Oliver has, nonetheless, to comply. There is to be no evacuation for the Black communities sheltering in the hotel. As the killing escalates and provisions diminish, so too Paul has to use all his negotiating skills and call upon his many contacts to help the hotel community to survive. As the Tutsi armed forces begin to gain the upper hand, Colonel Oliver and his UN troops are able to return and eventually evacuate all the refugees to safety.

To comprehend the Rwandan Genocide in which this film's narrative unravels, we need to understand the historical context. As Willis Okech

Oyugi (2019[9]) explains, pre-colonisation, Rwanda was a composite of eight kingdoms and was socially organised into clans, the most important of which were the Hutu and the Tutsi. By the mid-18th century the Kingdom of Rwanda, ruled by a Tutsi clan, became the dominant kingdom extending to englobe the other kingdoms into a coalesced whole by the mid-19th century. Both Tutsi and Hutu were migratory Bantu groups establishing themselves in Rwanda over a half century period, 1000–1500 AD. Their slow migratory arrival changed the nature of the original territory from forest to agriculture, thereby displacing the original indigenous inhabitants the Twa (who removed themselves to the mountains), although evidence points to a co-existence in which the Tutsi, Hutu, and the Twa were 'supplementing each other and living in close proximity on the hilly Rwandan terrains all along.'[10] The distinction between Hutu and Tutsi was, at this time, one of class and not racial – as it would become under German colonial rule. The Tutsi were pastoralists, land-owners, who thereby established their hierarchical status over the Hutu farmers.[11] Whilst there were undoubtedly rifts and social inequalities between Tutsi and Hutu populations, caused by the Tutsi privileged situation (ruling land-owners who leased their land and cattle to the Hutu), nonetheless, over the first three centuries (1500–1800) a reasonable co-existence prevailed.[12] Furthermore, to a relative degree, Tutsi-Hutu class distinctions were fluid; prosperous peasant Hutu could become Tutsi (through ownership of cattle, for example). Equally, Tutsi who fell on hard times could find themselves reduced in social status and obliged to become Hutu.[13] In summary, the Tutsi-Hutu economy was a feudal one (whereby land, cattle, and crops were used as exchange value), in which resources were shared (water, food, etc.), and – most significantly – in which the Tutsi-led militarised defence of the borders against slave-traders was maintained.[14]

It would be too simplistic, however, to say that, in terms of the Doughnut, transgressions occurred only in relation to social justice and equity. By the 18th century, Tutsi leaders, having expanded their territories, gained ascendancy, and developed a very advanced administrative

---

9 Willis Okech Oyugi (2019) 'Historicizing ethnicity and slave-trade memories in colonial Africa: The cases of Rwanda and Northern Cameroon' (66–86) https://escholarship.org/uc/item/1c84c7r3 (accessed 15 October 2019).
10 Willis Okech Oyugi (2019: 72).
11 Willis Okech Oyugi (2019: 70).
12 Indeed, during that period, the greater dissent and rivalries were between Tutsi clans themselves, as various kings sought to assert ascendancy over others.
13 For an historic overview see www.refworld.org/docid/469f38d51e.html (accessed 23 October 2019).
14 www.newtimes.co.rw/section/read/74419 (accessed 15 October 2019).

system, including laws that further indentured the Hutu. Rwanda became a powerfully centralised and militaristic state that was also 'characterized by an increasing socio-economic and political divide that favoured the Tutsi over the Hutu identity.'[15] Civil strife increased and, with it, land disputes provoked by recurrent droughts (lack of water to resource the effects of land conversion) – all of which served to increase 'tensions that eventually morphed into the Tutsi and Hutu divide.'[16] Increasingly, 'Hutu identity became associated with an "outsider" status' because of their dependence on the elite Tutsi.[17] This condition was particularly marked during times of hardship when the Hutu became more or less indentured serfs to the Tutsi (obliged as they were to offer their labour in exchange for food or taxes[18]). This divide was the one exploited by the Germans when they colonised Rwanda, turning a class division into an ethnic, racial one. Furthermore (as we shall see later), this concept of 'outsider' status would, post-colonialism, be one that would be turned back against the Tutsi with dire consequences.

Here is a plotting of Rwanda's ecological process, pre-colonisation (bold indicates the incidences of boundary transgressions):

> *Pre-colonial process:* Tutsi & Hutu slow migration (1000–1500)**=>deterritorializing aboriginals (Twa)**=>gradual installing of a class-driven holocenic regime (land-owner Tutsi/farmer Hutu) establishing land conversion **(forestry loss** but agricultural gain: living sustenance)=>a viable feudal economy (land, crops and cattle as exchange value)=>**creating hierarchies (Tutsi rule)**=>networks that protect all (anti-slavery)=>**but increasing lack of political voice, social justice and equity (Hutu)**=>droughts=>economic hardship (Hutu)

15 Willis Okech Oyugi (2019: 73) https://escholarship.org/uc/item/1c84c7r3 (accessed 15 October 2019).
16 Willis Okech Oyugi (2019: 72) https://escholarship.org/uc/item/1c84c7r3 (accessed 15 October 2019).
    As Oyugi explains: because the land and cattle-owning (and leasing) Tutsi had political superiority over the Hutu, 'pastoralists were advantaged over farmers (especially) during periods of drought since they could easily move their cattle to pastures afar (thus ensuring) greater economic stability.' Farmers meantime were bound to their piece of leased land and leased cattle (if they had any) and so inevitably suffered during seasonal setbacks. By the late 18th and into the 19th centuries, these tensions continued to intensify between Tutsi and Hutu as 'scarcity of land during the expansion period, meant that farmers were at the mercy of the pastoralists.'
17 Willis Okech Oyugi (2019: 73) https://escholarship.org/uc/item/1c84c7r3 (accessed 15 October 2019).
18 Willis Okech Oyugi (2019: 73) https://escholarship.org/uc/item/1c84c7r3 (accessed 15 October 2019).

A moral imperative to REVOLT   85

We see how migration and hierarchical systems bring about these transgressions including the impact on the environment by successive droughts due to land conversion. Let us now plot this process onto the Doughnut (Figure 4.1).

At this point, the ecology ceiling is not in danger. The slow migratory processes over a 500-year period did, though, lead to land conversion and cause the Twa to leave their hunter-gatherer land and to diminish as a population – for this reason the word 'land' in the Doughnut is in red. The shift to pastoral-agricultural practice did mean, though, that there could be a 'sustainable life' for all, as indicated in bold in the Green Biosphere of the Doughnut, except of course in times of drought. The consequences of land-ownership and class hierarchies – their socio-economic fluidity notwithstanding – did, indubitably, bring about social and political inequities, thereby transgressing the social

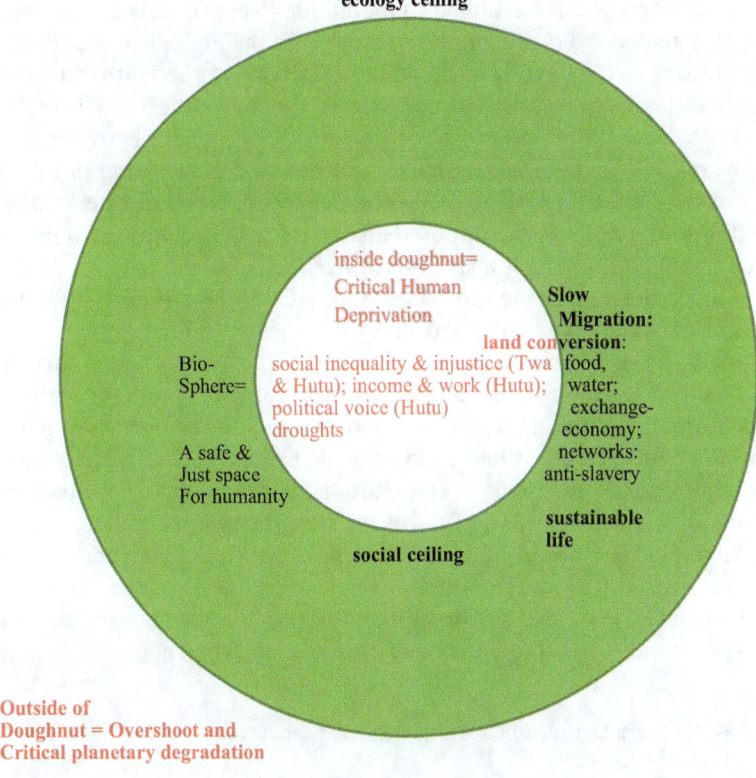

Figure 4.1 Mapping Rwanda's pre-colonial ecology.

ceiling in a number of domains (as indicated in red). As Raworth has constantly pointed out, ownership (who owns what and in what way, exploitative or extractive, as opposed to regenerative) is a key factor in transgressions of the biospheric boundaries. But the picture that emerges is not all in shortfall of the social ceiling. Indeed, during this period, a number of the 12 dimensions of the social foundation were safeguarded (food, water, networks combatting slavery, and an economy based on exchange value).

What now of the colonial process? The Berlin Conference 1884–5 (also known as the Congo Conference and the West Africa Conference) formalised European colonisation and trade in Africa. The Berlin Treaty 1885, signed by 13 nations of Europe and the USA, brought to an end the unbridled scramble by European nations to lay claim to African countries by putting in place the colonial partitioning of the African continent. The Kingdom of Rwanda was ceded to the Germans. Colonisation had a double, but ultimately, contradictory impact where the Tutsi and Hutu hierarchies were concerned. At first, the Germans – who ruled indirectly through the pre-existing political structure – reified the Tutsi as racially superior to the Hutu.[19] Under the rubric of science, the Germans determined that the Tutsi were descendants of the Hamitic-Ethiopian and therefore closer to the European ideal and more civilised. The effect of hierarchising in racial terms was to set in motion a division between the two (former) clans, a divide and rule system imposed also so as to deflect anti-colonialist sentiment and project it instead onto the Tutsi elite.[20]

Under Belgium's rule, post-World War One, the pre-existing Tutsi political system was curtailed and its own institutional system was implemented. Crucially, the Belgian colonial administrators further reinforced Germany's earlier division between Tutsi and Hutu ethnic identities by issuing ID cards which prevented any further movement between the classes, as had been possible before (these ID cards were not abolished until 1996[21]). The Hutu were designated as the indigenous Bantu people and the Tutsi as the Hamitic non-indigenous elite – a labelling that would have consequences post-independence, when, in a reversal of status, the Tutsi would become the ones designated as 'outsiders.' This racial distinction and fixing became entrenched and institutionalised through the education system. The Tutsi were taught

---

19 Willis Okech Oyugi (2019: 74) https://escholarship.org/uc/item/1c84c7r3 (accessed 15 October 2019).
20 www.sahistory.org.za/place/rwanda (accessed 23 October 2019).
21 www.monitor.upeace.org/archive.cfm?id_article=707 (accessed 11 November 2019).

in French and destined for white-collar work; the Hutu were taught in Swahili which was deemed 'sufficient for blue-collar jobs.'[22] To solidify these distinctions, Belgium enlisted the Catholic Church and missionaries to 'indoctrinate people,' especially the 'elites towards a European disposition.'[23] This doubling-up of racial distinctions undoubtedly fuelled mounting tensions and violence between the two clans. But, as Oyugi points out, two of the other major precursors to the genocide were, first, the gradual shift in economic power – thanks to the introduction of coffee plantation in 1904 by the Germans, which favoured the Hutu farmer by transiting the economy from that of an exchange to a monetised one – and, second, the publication of the Bahutu Manifesto in 1957, asserting Hutu-Power. These two factors – which were slow in their evolutionary impact – are closely inter-linked, as follows.

Under the effects of colonisation, which introduced coffee plantation, Rwanda's economy, by the time of independence in 1962, had become heavily reliant on its coffee crops that were produced by Hutu smallholders (during the 1970s coffee crops represented 60%–80% of Rwanda's export revenue[24]). The impact of a plantation economy, however, meant that, by the late 1970s, over 50% of agricultural land was lost to coffee (and later tea) production. Coffee production was controlled by the government (during and post-colonialism); they fixed the market price and the taxes levied. During the colonial period, it was the Tutsi as land-owners who were responsible for collecting the taxes. Post-independence, this land was re-appropriated and placed under Hutu governmental control.[25] Also post-independence, however, the population of Rwanda exploded, causing land shortages to reach a critical point and poverty to be on the increase. In the 1950s, the Rwandan population was around 2m, by 1990 it had reached 7m. Rwanda had become the most densely populated country in the African continent. A further contributing factor to this increased malaise and shortage was the economic crisis of the late 1980s, especially the crash of coffee prices in 1989, which affected all classes but most particularly the youth classes for whom prospects were slim. By 1990,

---

22 Willis Okech Oyugi (2019: 75) https://escholarship.org/uc/item/1c84c7r3 (accessed 15 October 2019).
23 www.sahistory.org.za/place/rwanda (accessed 23 October 2019).
24 http://siteresources.worldbank.org/AFRICAEXT/Resources/258643-1271798012256/rwanda_Coffee.pdf (accessed 24 October 2019).
25 http://siteresources.worldbank.org/AFRICAEXT/Resources/258643-1271798012256/rwanda_Coffee.pdf (accessed 24 October 2019).

as the Civil War gathered pace, the government had to increase resources to its army. It was forced to lower coffee prices further still. But it managed to divert 'its own economic policies onto the Tutsi/ RPF threat' and demonised 'the invaders, arguing that allowing Tutsis (currently in enforced exile) into the country would lead to Hutus having less access to already scarce land, and used the media to foment ethnic hatred.'[26] By 1993, anti-Tutsi feelings were running high, and the Hutu regime exploited the disaffected youth classes as a major recruiting ground for the two government-backed youth militia groups the Interahawme and the Impuzamugambi, the main perpetrators, along with the Rwandan Armed Forces, of the 1994 genocide.

For its part, the 1957 Bahutu Manifesto set the tone for the future nationalist Hutu movement, basing its rhetoric in the institutionalised racism imposed by the Belgian colonisers. The actual title of the Manifesto, 'Note on the social aspect of the indigenous racial problem in Rwanda,' makes this racialisation of the problem clear. The manifesto denounces the exploitation of the Hutu by the Tutsi and calls on Belgium to honour its post-war promise to bring Rwanda into independence.[27] One of the Manifesto's authors was an influential Hutu and the head of a coffee co-operative, Grégoire Kayibanda (future first president of post-colonial Rwanda). 'He and others awakened Hutu consciousness and denounced Tutsi political and cultural monopoly.'[28] Furthermore, in that rhetoric of Hutu-Power, the labelling of 'outsider' was now turned about and attached to the Tutsi, labelling them as the Tutsi Hamitic Invaders. This political rhetoric then morphed into the Rwandan Revolution (1959–61) – which was backed by the Belgian colonial powers (and the Catholic Church) whose mounting concern at the Tutsi's own agitation for independence made them fear for their control over their territories. This period of violence resulted in 'ethnic-driven killings,' some 10,000 Tutsi perished, and led

---

26 Karol C. Boudreaux (2010: 187) 'Economic liberalization in Rwanda's coffee sector: A Brew for success,' http://siteresources.worldbank.org/AFRICAEXT/Resources/258643-1271798012256/Rwanda-coffee-ll.pdf (accessed 24 October 2019) (parenthesis mine).
27 https://fr.wikipedia.org/wiki/Manifeste_des_Bahutu (accessed 24 October 2019). The original title of the manifesto is: 'Note sur l'aspect social du problème racial indigène au Rwanda' (Note on the social aspect of the indigenous racial problem in Rwanda).
28 Willis Okech Oyugi (2019: 76) https://escholarship.org/uc/item/1c84c7r3 (accessed 15 October 2019).

'to hundreds and thousands of Tutsi minorities being driven out of Rwanda.'[29]

During this period of exile in Uganda and the Congo (then Zaire) the Tutsi youth, in an endeavour to effect a return to Rwanda, formed themselves into an armed movement and called themselves the Inyenzi (meaning Rebel, Blue Monkey, and Cockroach). The Tutsi chose this term because, to their way of thinking, the cockroach was an animal that could easily invisibilise itself and gain entrance without being seen (again, this labelling would later be turned around against the Tutsi).[30] The Tutsi's rebellious assault failed, however. Thereafter, from 1961 to 1990 there were numerous armed clashes between the Tutsi and the Hutu – Tutsi attacks being repelled with the help of Belgian paratroopers and aided by the French military training and arming of the Hutu.[31] By 1990, the Tutsi exiles had reorganised themselves into the Rwandan Patriotic Front-Inkotanyi (RPF) – the latter epithet, Inkotanyi, meaning fearless.[32] A series of invasions launched by the RPF led to the signing of a new Constitution, in 1991, allowing for multi-party politics. A brief period of cease-fire ensued before being broken once again in February 1993.

A further treaty of cease-fire was agreed with the Arusha Accords Treaty of August 1993. But hard-core Hutu political elements attacked their Tutsi political opponents. The youth militia targeted Tutsi and Hutu moderates with beatings and killings. Thereupon, a plot was mounted by Colonel Théoneste Bagosora, Chief of Staff of the

---

29 Willis Okech Oyugi (2019: 75–7) https://escholarship.org/uc/item/1c84c7r3 (accessed 15 October 2019).
   This reversal of the Tutsi hierarchy was also one favoured by the Catholic Church; missionaries championed the rights of the oppressed Hutu people.
30 See Kennedy Ndahiro 'Dehumanisation: How Tutsis were reduced to cockroaches, snakes to be killed,' www.newtimes.co.rw/section/read/73836 (accessed 17 October 2019).
31 One instance of the French involvement in this terrible Civil War was under the code name of Opération Insecticide, a military operation organised by the French in response to the Rwandan government's request to train up an elite corps of the Hutu Presidential Guard who, in turn, were heavily involved in the genocide from the first day. The French also supplied arms. https://fr.wikipedia.org/wiki/Opération_insecticide (accessed 23 October 2019).
32 Although, as an indication of how little had changed in terms of the foreign military's position on the RPF, the Belgians for their part continued to support the Hutu forces as did the French military in situ who referred to the RPF as the 'Khmers Rouges of Africa.' Furthermore, both the Belgian and the French governments continued to support the Rwandan government financially. See www.refworld.org/docid/469f38d51e.html (accessed 23 October 2019).

Ministry of Defence, to accelerate the massacre of the Tutsi (he it was who established the Interahamwe, arming them with guns and machetes).[33] He was a hard-liner Hutu who refused to support the Arusha Accords. In order to provoke a mass public outrage and mobilise the army and the militia groups to eliminate those on his established lists of names (which included the Tutsi and moderate Hutu elites as well as ordinary citizens considered the enemy), he organised the assassination of the then President of Rwanda (Juvénal Habyarimana) and spread the claim that it was the Tutsi who were responsible.[34] The date was 6th April 1994, the beginning of the Rwanda Genocide. Meantime the UN Forces were rendered totally ineffectual.

Here's the plotting of this colonial to post-colonial process:

Colonial and Post-colonial process:

1885–1962: German colonisation from 'Treaty'=>land grab, racism, land conversion (to coffee production & shift of economy from feudal to monetary)=> social inequality perdures for Hutu =>1918=>Belgian colonisation from spoils of war=> racism, regulation through Belgian institutions=>social inequality, loss of political voice, injustice, but reversal of economic hierarchy for the Hutu=>**progressive agricultural land scarcity (given over to coffee production) and increased Hutu hostility (post 1957) leads to Hutu-Power de-territorializing much of Tutsi population (exodus of two thirds)**

1962–1994: Rwanda independence under Hutu regime =>market economy (coffee waxes and wanes)=>poverty & land shortages intensified by land scarcity (coffee and tea plantations)=>successive periods of civil war =>Three crisis phases: 1962; 1967–1973; 1973–1990=>leading to 1990 assault by RPF on the Hutu regime followed by successive conflicts between Hutu and Tutsi culminating

---

33 David Severnay in two articles in *Le Monde* (12 March 2019 and 13 March 2019) exposes the fact that the Rwandan military managed to get hold of arms (from the French) despite a UN embargo, and that Colonel Bagosora was at the heart of this deceptive move. www.lemonde.fr/afrique/article/2019/03/12/enquete-sur-le-financement-du-genocide-rwandais-felicien-kabuga-le-grand-argentier-des-massacres_5434679_3212.html and www.lemonde.fr/afrique/article/2019/03/13/enquete-sur-le-financement-du-genocide-rwandais-les-banquiers-les-dollars-et-les-armes_5435195_3212.html both (accessed 30 October 2019).

34 The air strike missile that brought down the plane carrying President Habyarimana (and his colleague President Ntaryamira of Burundi) was identified as French. The attempt to disguise this as a Tutsi attack was unfounded therefore since the missile was part of the Rwandan Army's arsenal, controlled by Bagosora.

A moral imperative to REVOLT 91

in the 1994 genocide[35]=>which included acts of gender inequality (rape); endangering health (deliberate infection of women with HIV); and the failure of UN or the West to commit adequately to ensure peace and justice

and here it is transcribed onto the Doughnut (Figure 4.2).

If we compare the two Doughnut models above (see Figures 4.1 and 4.2 together) we can visualise and determine the difference in terms of ecological impacts between these two periods – pre-colonisation (1000–1885) and colonisation and post-colonialism (1885–1994). A first evidence is just how much, from colonisation onwards, the Rwandan

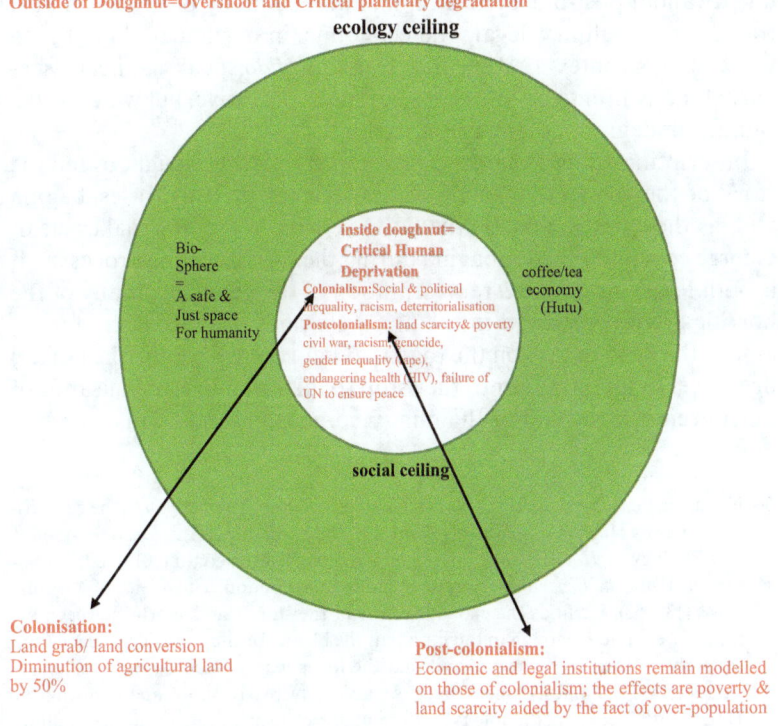

Figure 4.2 Mapping of Rwanda's colonial and post-colonial ecology.

35 France played an infamous role during this armed conflict period of 1990–4. It provided the Hutu regime with both arms and military training, in particular of the Interahamwe, the youth militia primarily involved in the genocide operations of 1994.

ecology (meaning also the population's well-being) is outside of the just and safe place for humanity. The shift from the lengthy period of migration into the development of a feudal system pre-colonisation was a gradual one; inequities notwithstanding it was a sustainable life for all. The immediacy of the colonialist land grab (as opposed to the slow migration of the pre-colonial period) was swiftly followed by land conversion converting half the agricultural territory over to coffee production – which subsequently, post-colonisation in the late 1970s, included tea plantations. Social rifts now became exacerbated into racialised antagonisms sparking the relentless, century-long violence stretching through colonisation to post-colonialism, the former bleeding into the latter not just in racial, but also institutional terms – for, Rwanda, post-independence, maintained many of the colonialist frameworks, military, legal, and economic, in particular. And it is to this complex context that the film *Hotel Rwanda* speaks. A heavy scenario for any film to countenance within its narrative, but which – to a significant degree – it managed to achieve.[36]

In scripting *Hotel Rwanda*, director Terry George sought to tell the story of the genocide through the experience of Paul Rusesabagina and, as the commentary on the DVD version of the film makes clear, George consulted Rusesabagina during the script-writing process.[37] It is both Paul's story and *a* representation of the genocide. Many of the horrific events portrayed in the film were ones that occurred: for example, the dead bodies on the road that Paul and Grégoire (the porter) drive over in ignorance until they stop to take a look; the thousands of Tutsi fleeing at the end of the film (George employed extras who had

---

36 Not all critics agree with this view. In particular, I'd like to refer you to the ecocritical article by Harri Kilpi on *Hotel Rwanda*: 'The Landscape's Lie: Class, Economy, and Ecology in *Hotel Rwanda*,' In: Paula Willoquet-Maricondi (2013) (ed), *Framing the World in Ecocriticism*, Charlotteville and London, University of Virginia Press (135–53). He finds the ecological side of the story (land shortage especially) is not fore-fronted enough in the film; that the blame for the genocide is placed too squarely on the Interahawme, and that the focus is too much on the two central characters: Paul and Colonel Oliver. As my analysis will show, and by using the Doughnut Theory methodology, all the major elements are evoked in the film's narrative. Or again, an article by Okaka Opio Dokotum (2013) 'Re-membering the Tutsi Genocide in *Hotel Rwanda* (2004): Implications for Peace and Reconciliation,' which argues that, despite its powerful creation of awareness about the Tutsi genocide, *Hotel Rwanda* contributes to a postgenocide conflict and even undermines justice, peace and reconciliation' In: www.jstor.org/stable/10.2979/africonfpeacrevi.3.2.129?seq=1#page_scan_tab_contents (accessed 01 November 2019). At least these dissenting views reveal the power of the film to raise questions.
37 *Hotel Rwanda*, DVD, Kigali Releasing Ltd (2004).

actually experienced this terror); and mention must be made of the evacuation of the Whites onto the bus and the quite grotesque image of the woman with her dog sat at the window with her husband standing behind her complacently taking a photograph (as if a holiday snap) of the hotel and the abandoned Black communities in the foreground. In most instances, George reconstructed the events; but on one crucial event – the one in which the journalist brings back evidential footage of the massacring of the Tutsi by Hutus armed with machetes – is the actual footage.[38]

There is, then, a strong documentary feel to this film. Feeding into this style is the presence of the radio station RTML which we hear for the very first time as the film credits open the film. The role played by RTML in the lead up to and during the genocide has been widely acknowledged.[39] Radio was *the* means of communication in Rwanda, for Rwandans listened to their radios for their information, and RTML was particularly popular with youth audiences because of its mixture of popular music (often Zairean) and comedy as well as its 'flash' headlines (worthy of today's Tweets). RTML was explicitly a Hutu-Power broadcasting company (founded and bankrolled by Félicien Kabuga and indirectly financed by the government through its sister channel Radio Rwanda).[40] A RTML radio broadcast opens the film, immediately focussing our attention on the critical situation in Rwanda and making clear the importance of this Hutu-Power radio station as an instrument of propaganda and as agitator (if not instigator), through its rhetoric of hatred, of the campaign of extermination of the Tutsi. The opening words set the tone: 'this is our land,' 'the Tutsi have taken our land,' 'they are traitors and invaders,' 'they are cockroaches and we must exterminate them.' Later, after President Habyarimana is assassinated, RTML declares he 'was murdered by the Tutsi cockroaches.' We note how the Hutu rhetoric has incorporated

---

38 Nick Hughes was the journalist who filmed the footage. He speaks out in 'Exhibit 467: Genocide through a camera lens,' www.internews.org/sites/default/files/resources/TheMedia&TheRwandaGenocide.pdf (pages 231–34) (accessed 28 October 2019). So too does Mark Doyle 'Reporting the Genocide' in the same publication (pages 145–59).
39 See https://en.wikipedia.org/wiki/Radio_Télévision_Libre_des_Mille_Collines (accessed 30 October 2019).
40 See https://en.wikipedia.org/wiki/Radio_Télévision_Libre_des_Mille_Collines (accessed 30 October 2019). Kabuga is still wanted by the International Criminal Tribunal for Rwanda. He also financed the purchase of 500,000 machetes in 1993 (evidence of which is pointed to in the film with the case of machetes imported from China by Rutaganda).

a number of references to earlier periods in the Rwandan history: the claim that their land has been stolen and that the Tutsi are invaders, not only positions the latter as the outsider who has exploited the Hutu, it also harks back to the colonial racist division of the two clans into racialised ethnicities and to their labelling the Tutsi as the Hamitic non-indigenous 'outsider.' The term cockroach, formerly, in Tutsi consciousness, a word intended to designate their rebellious courage against their forced departure from Rwanda in the 1957–62 period, is now picked up by the Hutu propaganda machine and turned against them as a term of abuse. And it is RTML that launches the code word for the genocide to begin: 'cut the tall trees' (a reference to the Hamitic origin of the tall Tutsi). All of this rhetoric also makes clear the Hutu's long-held resentment of their subaltern status, culminating, first, in a reframing of their resentment into systemic racism, and, second, into a determination to exterminate all cockroaches – that is all Tutsi and all Hutu sympathisers ('Hutus who shelter Tutsi cockroaches are all cockroaches,' declares the radio).

Language in this film has a determining role in other contexts as well. Paul is an embodiment of the 'Rwandan Way' that is used to solve difficulties, namely, 'to sit down and negotiate.' It is thanks to his expertise in this domain of language and its effects that he manages to preserve the lives of refugees in his hotel. At first, he is able to sit down with his 'opponents,' such as the local leader of the Interahamwe, George Rutaganda to obtain supplies for the refugees; he also uses it successfully in his interchanges with General Augustin Bizimungu to ensure some protection for those taking refuge. Later, after the UN has abandoned him and his people, and as the situation reaches crisis point, Paul has nothing more to offer in terms of gifts, so General Bizimungu also withdraws support. At this desperate juncture, Paul uses his language to persuade Bizimungu that he must protect them, if only to save himself once the horror is over. 'We need to help one another' Paul states, adding 'the Americans have you on a list of war criminals, they'll be coming after you. You need *me* to tell *them* how you helped us.' His moral rhetoric works, Bizimungu orders the Interahamwe and his army troops out of the hotel.

In the face of the terrible genocide of 1994 and the gross neglect of the Tutsi by outside forces, there are stories of acts of humanity in *Hotel Rwanda*. Intriguingly, if we were to take a genre-based approach to this film, we might be inclined, as a number of critics at the time were, to see *Hotel Rwanda* as a run of the mill disaster movie which merely picks up on the sensationalised moment of the genocide and leaves the historical causes in the realms of silence, preferring instead to highlight the courage of two men (one Black – Paul the hotel manager, the other White – Colonel Oliver the UN chief representative).[41] A Doughnut approach shows, however, that this is not the case. Moreover, on a mere narrative level, we witness numerous acts of kindness and courage, amongst others, by hotel staff (who tend to the children, for example), Paul's wife whose moral compass enables Paul's own growing humanity, the Red Cross workers – all of whom are risking their lives by helping the refugees. What emerges is the mixity of human responses when caught in critical adversity (thus, the porter Grégoire's hatred of the Tutsi leads him to a series of infamous denunciations, but his fellow colleague, Dube – a Tutsi – makes all efforts to ensure the safety of the people besieged in the hotel).

The way in which the genocide is represented in the film is measured, documentary in style and not sensationalised. This was a deliberate choice, according to Terry George. To this effect, he filmed the scenes of massacre in long shot – electing not to recreate the full brutality.[42] The point was to show the extent of the horror but not to the degree that it became normalised (as violence in movies has become, with spectators fairly inured to the spectacularisation of acts of horror). As one of the journalists says, the risk is that 'people will say "how horrible" and will go on eating their dinner.' To that effect of understatement, there are numerous instances of it within the film. The Red Cross representative, Pat Archer, who explains that she has brought the Tutsi orphans to the hotel to save them from their systematic murder by the Hutu who want to eliminate future Tutsi generations (*stating*, rather than *showing* the killing is enough). From the brief *glimpse* we get of Tutsi women imprisoned in cages at George Rutaganda's compound, it is clear from the soundtrack that they are being repeatedly raped, showing, *elliptically*, the intention to infect the women with HIV so that no future Tutsi children will survive.[43] Paul being obliged to produce his ID card to confirm his status, even though the military are fully aware he is Hutu, serves to *underline* the precarious nature of his wife's Tutsi status (already menacingly hinted

---

41 See Kilpi (2013) and Dokotum (2013).
42 Terry George in the DVD commentary. He explained that he relied on this kind of footage to tell us what we already knew. Incidentally, this kind of elliptical shooting of horrendous events is one that Jean Renoir adopted in many of his 1930s films.
43 The systematic rape of Tutsi women by troops and militia was invoked as a duty by (of all people) the then Minister for Family Welfare and the Advancement of Woman, Pauline Nyiramasuhuko. She was later convicted of crimes against humanity.

96 *A moral imperative to REVOLT*

at several times by George Rutaganda). The blood-stained UN blue helmet delivered by the Hutu Impuzamugambi militia-man adorned with a purple wig and wielding the electric drill he used to kill the soldier – acts as the *warning signal* that the UN will be forced to leave. Finally, the presence of the machetes, for the most part *heard* on the sound track as a metallic striking and slicing, rather than *seen*. Ellipsis, synecdoche, allusion – all are powerful methods of representation.

As we transcribe the film's narrative onto the Doughnut model, we can see how the two temporalities (past and present) feed into each other: the historical context and its slow evolving violence alongside with the contemporary rapidly changing socio-political reality which hurtles along at an unconscionable speed (Figure 4.3).

In the inner circle, the red asterisk denotes where the narrative starts (with the loss of land being mentioned on RTML), and the red arrow points to the history that the film refers to, namely, the effects of colonialism (racism, inequality, and injustice). The black lines indicate the causes of ecological overshoot and their impact. As Raworth's model

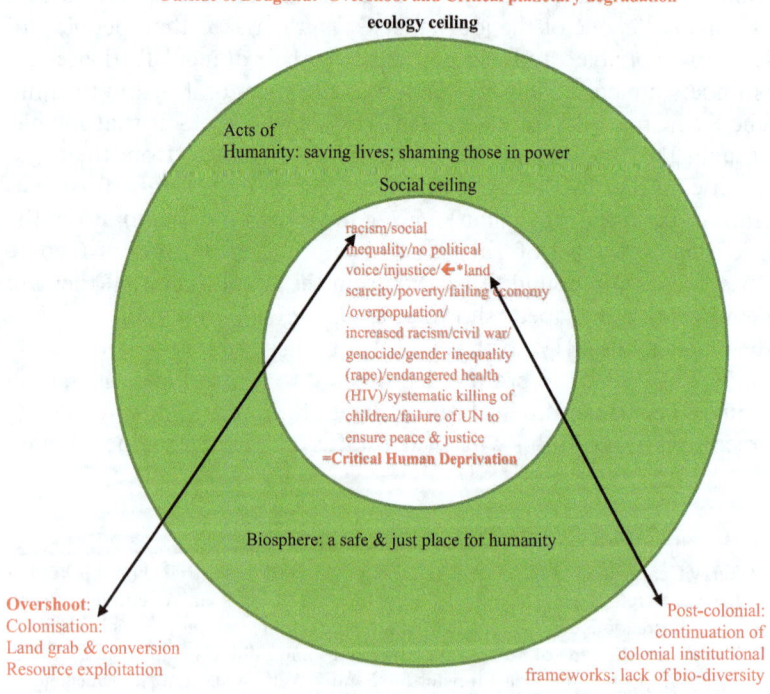

Figure 4.3 Mapping *Hotel Rwanda*.

makes clear, both ceilings are transgressed, leaving only *one* element (acts of humanity) within the safe and just space of the biosphere. Furthermore, Intersectionality is fully in evidence in this film in that it shows history at work, illuminating the «*how*» and «*why*» of events – revealing how deep are the wounds of racism, bound as it is into privilege. For, if pre-colonisation, the distinction between the two clans was one of class, and one which had a relative fluidity, under colonialism that distinction was turned into racial differentiation and racial difference, in turn, became a marker of privilege – the Tutsi elite. The irony of course is that, post-colonialism, the elite, particularly the political elite, now primarily emanated from the Hutu (with a Tutsi elite class very much in the minority). But, because conditions did not change for the struggling agricultural working-classes (because of a lack of biodiversity) the old animosity against a perceived Tutsi privilege prevailed. In this context of lack of biodiversity, it is worth noting that, once the coffee and tea production had absorbed half of the agricultural land, the share of food crops never exceeded 6.5% and that, of the total food produced, only 34% reached the markets. The reasons for this poor capacity lies, first, with the lack of appropriate technologies and, second with the lack of an adequate water supply.[44]

Hotel Rwanda reveals the transgression of ecological and social boundaries which is at the heart of this genocide and which has historical roots. This is clearly plotted throughout the film. The impact of the 80 years of colonisation is made explicit (by the journalist in the bar), as is the fact that colonial interests still function both as part of the economy (the Belgian company Sabena) and as a military presence (Belgian and French[45]). Equally, the enduring Civil War between Hutu and Tutsi, post-1962 is referred to (not least by Paul when we learn how, long before he was married to Tatiana, he rescued her from the northern province where she was in danger from Hutu violence). The diverse systemic racisms are clearly evoked (that of the colonialists,

---

44 See www.fao.org/rwanda/fao-in-rwanda/rwanda-at-a-glance/en/ (accessed 14 November 2019). We should also recall the words of Acemoglu and Robinson about the 'uneven dissemination of industrial technology and manufacturing production.' (2013: 51).

45 A further instance of the French involvement in this terrible Civil War (other than Opération Insecticide, see footnote 31), that further reveals their collusion with the Hutu position, was Opération Turquoise which, under the guise of a humanitarian military exercise to put an end to the genocide, in fact allowed for the exfiltration to the Congo (then Zaire) of numerous perpetrators of the Rwandan genocide. www.lalibre.be/dernieres-depeches/afp/gacaca-inyenzi-les-mots-du-genocide-rwandais-574c36f135702a22d7db67fa (accessed 23 October 2019).

but also that between the Tutsi and Hutu; and that of the West in its refusal to intervene, only coming to rescue the Whites caught up in this Civil War) – as Paul rightly states, ensnared as he is in the crossfire of all these racisms, 'I have no history' – that is, as a Black African, I do not matter to the outside world, I am a non-being. Finally, in this film, there is no shortage of footage of the different strategies used by the various factions of the Hutu military, the militia groups, and the self-appointed Hutu civilian executioners, to degrade, humiliate, and exterminate the Tutsi. Against this systemic racism and conflict of interests, there is no redress, only, as Paul says, a last ditch revolt of shaming those in power to help – and even this is embedded in another effect of post-colonialism: the creation of 'elites' first imposed by the Belgians. This social hierarchisation comes into play when the elites (Hutu and Tutsi) sheltering in the hotel manage to call in favours from contacts abroad, however, only they get to be evacuated – the rest are left behind to face their indubitable fate.

Privilege and the economic, political, legal triangulation I spoke of earlier (in Chapter 3), point to a circularity of self-interest that is manifest in this film. The film itself feels cyclical. On three occasions Paul circles from his hotel, to George Rutabanga's emporium and back to the hotel, to negotiate both for provisions for his refugees and bargaining power (for example cases of beer which are as valuable 'as gold'); on three occasions he interacts with General Bizimungu to protect his hotel refugees; on three occasions the UN comes to the hotel to effectuate peace-keeping (and failing) and people evacuations (one failure, two successes). So, indeed, the film narrative has a circular shape (in three ever-decreasing circles, one might say), rather than a linear one, making it a far cry from the traditional Hollywood product of a three-act arc narrative. As I hope the above analysis shows, despite what some critics such as Harri Kilpi believe, the narrative does *not* follow in 'the conventions of Hollywood storytelling bias' (2019: 138), namely, focussing on the 'extraordinary' (145), pitting 'good against evil' (149).[46] For, there is *no* concocted Happy Ending to this film. That the refugees do eventually get evacuated, that Tatiana finds her two orphaned nieces in the Tanzanian refugee camp – to which she and the rest of hotel inhabitants get taken – is a fact, not a contrived resolution. Paul is not 'extraordinary.' Yes, he is middle-class (an educated

---

46 Harri Kilpi 'The Landscape's Lie: Class, Economy, and Ecology in *Hotel Rwanda*,' In: Paula Willoquet-Maricondi (2013) (ed), *Framing the World in Ecocriticism*, Charlotteville and London, University of Virginia Press (2010: 135–53).

Hutu, thank you Belgium!) and serves in an hotel that typically receives the elite classes (amongst whom he moves with ease, thanks to his acquired skills of negotiation); but in that middle-classness that he inhabits, he is ordinary – and he develops his moral courage over the course of the story. The mixity of motivations – deeply entrenched in some cases – is made clear through the various characterisations (also based in real people), and the issue of good versus evil is less at the centre of this film than the cruelties that hardship, on the one hand, and political ambition, on the other, impel people towards. Never forget Hannah Arendt's famous words: it is 'the banality of evil' that shows its face most easily in times of extreme crisis.

The imperative is to take a stand, to revolt with the means to hand – in *Hotel Rwanda*'s case, this was a story that had to be told (as director Terry George stated[47]), a story that had to be told because 'the world let it happen' – words that are echoed in the BAFTA-awarded 2019 documentary about the Syrian Civil War, *For Sama,* filmed by Waad Al-Kateab between 2011–16 in Aleppo, starting with the civilian protests against President Assad, then the siege and violent attacks by the Syrian regime, aided by the Russians, to crush the city's population. For here, too, the world let it happen, by doing nothing.

The need to do something, in this instance to draw people's attention to child poverty, is also what propelled Nadine Labaki to make the film *Capharnaüm/Chaos*, perhaps a film less familiar to the reader, and the only non-English-speaking film to be addressed in this book. Told from a 12-year-old child's point of view, Zain, the film points to the many criss-crossing of causes to this destitution, of which there are three dominant ones: first, conflicts and wars, causing waves of refugee migration over the past 50 years into Lebanon, beginning with Palestinian refugees from various historical conflicts (originally three quarters of a million over a 20-year period, and now currently estimated at 174,000), some 8,000 Iraqi and Sudanese refugees, and, most recently the massive influx of Syrian refugees (1.5m); second, economic migrant labour, chiefly from Ethiopia (120,000 or so, many not legalised and bought very cheaply); third, systems and governments that pay no heed to the slum life right on their doorstep. Presently, refugees (about 1.8m) and displaced persons (around half a million)

---

47 Terry George quoted in Kilpi (2019: 137). On this subject of telling the story of the Rwandan genocide, first a very interesting TV series *Black Earth Rising* (2018), deals with the subject from the point of view of the trauma in the aftermath of the genocide (BBC Two); and a French documentary on the French involvement *Retour à Kigali*, Jean-Christophe Klotz, 2019.

make up 33.8% of Lebanon's population (2.3m out of 6.8m).[48] Around a quarter of the Lebanese, themselves, live in poverty. Beirut alone has over half a million individuals living in illegal settlements, slums or shanty towns; the rest of the poor live piled up on top of each other in insalubrious buildings, paying rent to unscrupulous landlords. This is the world Zain inhabits.

And so as not to underestimate the film's importance, I thought it interesting to begin with the fact that this film cost $4m to make (and was five years in production) and garnered a huge world-wide box-office revenue of $68.6m, whereas *Hotel Rwanda*, for its part, was produced for $17.5m and made $36.5m at the box-office (including the international revenue of $13m). Just as interesting, the largest part of the market for *Capharnaüm* was China, at $54m (79% of the box-office revenue). Three factors contributed to this success: word of mouth, the fact the film was nominated for an Oscar, and the growing interest in China for art house cinema.[49] To which we should add the final element: the universality of the story, poverty, migration, what it means to be a social outcast, to be street children – for, all of these are relevant and global issues today, explains producer Mohamed Hefzy.[50] And, on the back of its Cannes Prize and Oscar nomination, the film was sponsored for a New York première screening by the UNHCR (the UN agency for refugees), an event which brought about positive outcomes for some of the cast (see below).

I shall be developing the narrative story in greater detail, but, for the moment, here is the briefest of synopsis to get us started. The film opens with Zain seeking justice against his feckless and unscrupulous

---

48 Latest available figures (23 March 2019) from: https://libnanews.com/liban-2285-millions-refugies-syriens-palestiniens/ (accessed 04 November 2019).
49 See review in Hollywood Reporter, Alex Ritman (2019) 'What the $50 million success of *Capernaum* in China means for Arab cinema,' www.hollywoodreporter.com/news/what-china-box-office-success-capernaum-means-arab-cinema-1218783 (accessed 01 November 2019). Incidentally, the winning film, that year, for the Oscar best foreign film category was the controversially nominated *Roma* by Alfonso Cuarón (2018) – controversial, because it was a Netflix production which technically did not meet the theatre release requirements for nomination. Whilst it cost $15m to produce, it only garnered a box-office of $5.1m, obviously, because of its limited theatre release! There is no trace of *Hotel Rwanda* being screened in China. But it is interesting to note that post-1994, China became one of the major investors in Rwanda's shifting economy, to electronic technologies amongst others.
50 Hefzy quoted in Alex Ritman (2019) 'What the $50 million success of *Capernaum* in China means for Arab cinema,' www.hollywoodreporter.com/news/what-china-box-office-success-capernaum-means-arab-cinema-1218783 (accessed 01 November 2019).

A moral imperative to REVOLT 101

parents for the 'crime' of giving him life. Through a series of flashbacks, we learn how it is that he has been jailed for the crime of stabbing Assaad, the local store owner and the son of his parents' landlord, for causing the death (through a miscarriage) of his beloved 11-year-old sister, Sahar (whom his parents had sold to Assaad in marriage to pay off their debts). To that end we follow Zain, a fearless streetwise child as he flees his abusive parents; survives through his wits on the streets; encounters an Ethiopian refugee Rahil and her baby son, Yonas, with whom for a brief moment he finds some comfort; and finally, when all seems to condemn him to perdition, manages through a television phone-in show (on injustices against children) to denounce his parents and bring them to answer in a courtroom.[51]

The cast of *Capharnaüm* was almost entirely non-professional (two major exceptions being the judge and Zain's lawyer, both actors). The street kids played themselves, as did the slum inhabitants and local storekeepers. Several of the lead roles were embodied by refugees or illegal immigrants. Zain (Zain Al Rafeea), a Syrian refugee in real life, had been on the streets of Beirut for five years (after the film, thanks to the efforts of UNHCR, he was moved with his family to Norway); Rahil (also known as Tigest, played by Yordanos Shiferaw) was an illegal economic migrant from Ethiopia (in fact she was arrested during the filming and jailed for two weeks before the producers could get her released); her son, Yonas (actually played by a baby girl, Boluwatife Treasure Bankolé), though born in Beirut of African refugee parents, was deported along with her mother back to Kenya and her father was sent back to Nigeria; Sahar (Cedra Izam), Zain's sister was a Lebanese street kid (thanks to UNICEF, she was enrolled into school, as was Maysoon, played by Farah Hasno, a Syrian refugee whom Zain befriends in the film).

The opening shots of the film locate the subject matter firmly in four major environments: the slums of Beirut (in a flying overhead-shot provided, presumably, by a drone camera), the young offenders' prison (where Zain, in his underpants, is being examined by a doctor to determine his age), a detention centre (in which illegal migrant women are being processed, including Rahil/Tigest), the court (into which Zain is brought, handcuffed, to present his civil claim against his parents). Each of these snapshot scenes will gain relevance as the film's narrative unfolds through a series of embedded flashbacks.

---

51 For a fuller synopsis see: https://m.the-numbers.com/movie/Cafarnaum-(Lebanon)-(2018)#tab=international (© 1997–2019 Nash Information Services, LLC.) (accessed 04 November 2019).

We visit the court on four occasions, each one functions as a springboard into the trajectory that has brought Zain to this place of reckoning. Each visit presents a moment of witnessing of past events. Thus, the first moment in court triggers the flashback that fills in on Zain's 'home' life, an environment in which his parents both abuse and exploit him. He is forced by them to work for the local store owner, Assaad, rather than be allowed to go to school (which he yearns for). He is coerced by them to obtain the drug Tramadol from the various nearby pharmacies by presenting forged prescriptions. This drug is then processed by Zain and Sahar under instruction from their mother and soaked into clothing which she subsequently delivers to her eldest son who is in prison (so he can then sell the drug onto inmates). Having noticed that his sister has started her periods, Zain tries to protect her from the fate awaiting all young girls – namely, being 'sold' to older men. He plans their escape, but to no avail. His brutal father apprehends the two before they can get away and drags Sahar off to her fateful nuptial with Assaad. It is this event that triggers Zain's decision to flee his parents – to which they react with fury, then indifference (they have enough children left in their insalubrious living space to carry on these above duties for them).

The second moment in court begins with Zain's father complaining about his dreadful life, presenting himself as the victim, rather than acknowledging any responsibility: 'I didn't want any of this, people despise me; I was born into this.' Adding, as a cock-a-snout to the judge, 'maybe I could have been better than you.' Indubitably, there is some truth in these words – poverty is an appalling condition. However, we have witnessed the father's awful brutality towards his many children, in particular Zain. We have seen him zoned out on dope, unwilling to lift a finger to help even his wife. So any prospects for betterment seem not realistically to be within his purview. Indeed, poverty, as shown in this film seems to induce two types of contrasting behaviour: exploitative cruelty, or compassionate acts of kindness. And, as we segue into this second flashback, it is this latter behaviour that comes to the fore. Zain encounters a strange old man dressed up as Spiderman. The old man tells him in fact he is really Spiderman's cousin, Cafardman (Cockroachman). Intrigued, Zain follows him to the local fair (Freij Fun Fair City), where Cafardman works as a parking attendant. There he meets Rahil, known at this stage under her false name of Tigest. She takes him under her wing and offers him shelter in her pitiful shanty home which somehow she has managed to render warm and welcoming.

Once again we are back in court. It is Rahil's turn to give her witness testimony, and we cut back to the third flashback which shows how, in exchange for shelter, Zain cares for her baby son, Yonas. A tender bond grows between all three. Rahil/Tigest, an illegal migrant, meantime is trying to secure new false documents from a so-named Aspro who, knowing she hasn't enough money, wants to exchange the documents for her son – obviously with the intention of selling the baby on for profit. Rahil/Tigest, at her wits end, turns to Spiderman/Cafardman who along with his trans woman-friend endeavours to help her gain legal documents. They fail. Rahil/Tigest sells her hair to raise the money, but too late. She is arrested as an illegal and taken to the holding cells we first witnessed at the beginning of the film. From this point on, Zain's existence once again enters into perilous waters, as does that of Yonas who only has Zain to protect him. For a while Zain manages. Encouraged by a newly acquired friend, Maysoon, a Syrian refugee girl, he endeavours to raise enough money (by resorting to the Tramadol trick he learnt from his mother and selling Tramadol water) to get himself and Yonas trafficked to Sweden. He loses the money and it becomes evident he cannot care for Yonas; his last hope is to save himself, and he sells Yonas to Aspro. To obtain the ID he needs to escape to Sweden, Aspro demands that Zain get his birth certificate. He goes back to his parents' place to retrieve the documents only to be told by his jeering father that none of the children have been registered. 'We are less than nothing' his father shouts as he assaults his son. It is at this point that the truth of Sahar's death through miscarriage occurs and Zain flips, grabs a knife, and goes after Assaad intending to kill him.

What is remarkable in these two particular flashbacks is the presence of the carnivalesque – the fair, its inhabitants, the diverse sexualities, the masquerades (no one is who they say they are). And yet it is a place of safety, at least up to the moment when Rahil is arrested. As Zain enters into the world of the fairground, the first thing he does is to clamber on top of a merry-go-round in the form of a hugely breasted woman and pull away at her blouse to reveal her bosom. It is this humorous act of Zain's defiance, observed by Rahil/Tigest, that marks the beginning of their connection. The fun-fairground becomes, then, not just a place of safety, but a place of access to the true maternal – embodied by Rahil, and, in a way too, by the huge-breasted fairground figure. Henceforth, until her arrest, Rahil cares for the two boys, Yonas and Zain, and in return Zain cares for Yonas – including carrying out the maternal duties of nappy-changing and bottle-feeding. Beyond that first circle of care, Cafardman and his trans friend do their best to make the trio safe and, even if their masquerade as a straight couple intent on employing Rahil fails, the humanity is there as is the willingness to risk arrest as fraudsters.

But it is also the case that, within these same two flashback sequences, there are strong contrasting tones between, on the one hand, the compassionate behaviour of those connected with the fairground – within the world of the carnivalesque, a chaotic, vertiginous world of the unreal where moral compasses prevail; and, on the other hand, the ruthless brutality outside that world, as embodied by Aspro whose only interest is in making money out of the misery of others: selling fake papers at inflated prices, dealing in human trafficking (selling babies and exploiting refugees and the poor trying to escape to a better life in Scandinavia) – the chaotic world of the real. A meeting, or rather clashing, of two worlds around the same issue: poverty.

We return to the court one last time. The witness taking the stand is Assaad, who survived the attack. He claims he had no idea how young Sahar was. The judge is deeply unconvinced. Zain's mother then pipes up protesting – if not her innocence in 'selling' her daughter on – that she has, however, been 'a slave all my life.' A sentiment echoed by Zain's father who shouts out 'we're the victims!' But Zain has made his case against his parents. And as the film draws to a close, we observe Aspro's trafficking scam being busted by the police; the police discovering Yonas in the mêlée and returning him to Rahil at the point of her being deported back to Ethiopia and, finally, Zain being photographed for his passport in readiness for his departure for Scandinavia.

Of the four major environments first seen in the opening establishing shots of the film, it is interesting to note, first of all, that three out of the four are in some way attached to the judiciary and penal correction. Thus, the court is the dispensary of justice; and the detention centre and young offenders' prison the province of punishment. However, in this world of chaos (the *Capharnaüm* of the film's title), these spaces are not, where Zain is concerned at least, places of oppression and silencing. Far from it. When Zain's age is being assessed, he is treated gently by the medical officer. When in prison, Zain is able to phone into the television programme on injustices to children and thereby gain access to the civil court. Over the phone-in to the television station, he denounces the life he has been forced to lead, in which his parents have treated him as 'a son of a bitch' and a 'piece of shit.' Meantime the entire young offender inmates cheer in support – knowing that this is the life they all lead, as indeed Labaki discovered. When doing research for the film, she interviewed street kids and many spoke of how they wished they were not alive, seeing themselves as insects and parasites, as not existing, being invisible, non-beings.[52]

---

[52] See: Bilal Qureshi's article 'In *Capernaum*, The chaos of Lebanon from a homeless child's perspective,' www.npr.org/2018/12/16/676553757/in-capernaum-the-chaos-of-lebanon-from-a-homeless-child-s-perspective?t=1573569175567 (accessed 11 October 2019).

A moral imperative to REVOLT 105

In court where Zain pleads his case, the judge listens attentively to all the evidence presented to him (including the protests of the parents), puts questions to the various witnesses and to Zain as plaintiff, and is seen to be exercising his power to assess the value of the testimonies with patience and equanimity. In fact, of the four environments, the only place of oppression and silencing within these domains of the judiciary and the penal is very clearly the detention centre in which illegal migrants are huddled up together in cages according to their sex. Even here, however, a brief note of uneasy levity is introduced when some Christian missionaries come in and play some jolly tunes – a fairly grotesque moment until the detainees actually join in by dancing, thereby momentarily entering the disruptive chaos of the burlesque.

So now, as we come to place all this information on the Doughnut, this is the picture that emerges (Figure 4.4).

Within the green of the biosphere there are instances of safe and just spaces for humanity: spaces of the maternal and familial offering

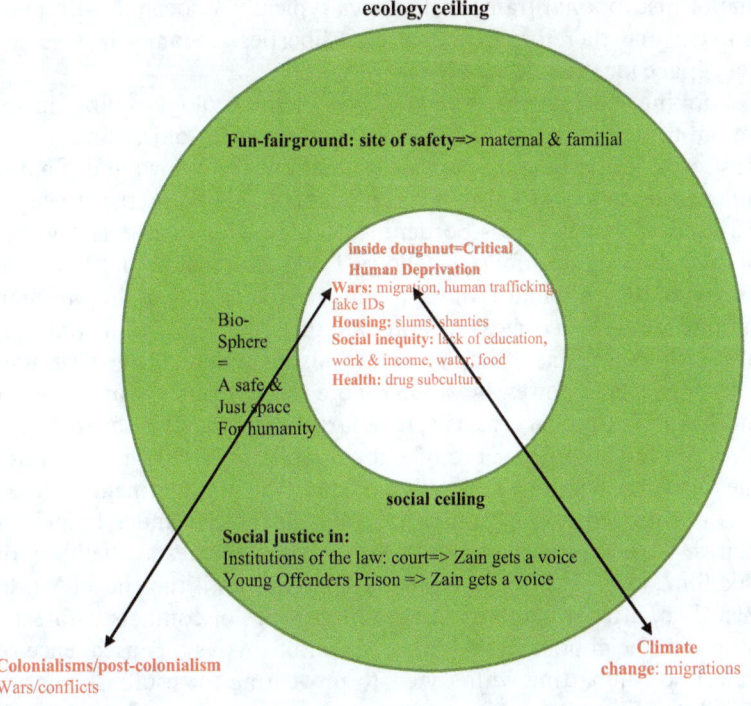

Figure 4.4 Mapping *Capharnaüm*.

respite for the abused (the fairground and Rahil's shanty home); and spaces of social justice offering the abused a place to voice their complaints in safety (the young offenders' prison and the court). Curiously, neither are spaces we would necessarily associate with safety or perhaps even justice. First, there is humanity emanating from the marginal space of unbridled chaos (the fun-fair, a place of vertigo, chance and carnival) in the form of Cafardman, his friend (name unknown), and Rahil/Tigest – all three also on the margins in one sense or another. Second, there is safety and justice dispensed within institutions of the law – not just in the instance of the court and the humane judge, but also the young offenders' prison, and in the closing moments of the film, the police enabling the return of Yonas to his mother.

Both the marginal and the institutional domains are shown here as working to counter the transgressive effects of overshooting the social ceiling, therefore. The first, the fun-fair, is arguably a non-space, a space of transience, unfixity – yet one where relations form, in this instance sub-cultural ones (gay, trans, illegal migrant, runaway), thereby creating a community that in its very *dasein* (beingness) exposes the failures of society and challenges its authority. The latter domain, the one of institutional frameworks, ones typically associated with punishment and therefore denial of civil liberties, becomes instead *the* very space for the voicing of human rights.

Looking, now, at the transgressions of the ecology ceiling, in relation to Lebanon, the impact of numerous colonisations on its post-colonial heritage of partial occupation (see below) and, finally, independence has meant that it is a country striated by the effects of wars, civil conflicts, cross-border invasions and finally differing waves of migration. This unfolds as follows. Long-term history has Lebanon as part of the Caliphate (7th–16th centuries), then part of the Ottoman Empire (16th–20th centuries); and even these periods are far from stable, with as just one example the various incursions of the Crusades causing shifts in power. Lebanon's more recent history, which is what more poignantly concerns us here, is largely embedded in the so-called «Middle-East Problem», brought about largely by what occurred after the Ottoman Empire was dismantled and the transcontinental region was partitioned into a number of countries placed under French or British Mandates. Lebanon and Syria, two countries at the heart of this film, were both under France's Mandate until the end of World War Two, after which they underwent a series of conflicts with each other and neighbouring Palestine (much of this as a consequence of Palestine combatting with a view to preventing the establishment of the State of Israel).

In the case of Lebanon, however, a complex picture emerged, leading to the Civil War 1975–1990. Initially, post-independence, Lebanon continued with its parliamentary structure put in place by the French and which favoured the Christian population. The advent into Lebanon of the various waves of Palestinian refugees during the 1948–67 period (overall about three-quarters of a million) led to a shift in demographics which favoured the Muslim population. Initial fighting between the Lebanese factions (Muslim and Christian) escalated into this long drawn-out 15-year Civil War into which Syria and Israel intervened to fight alongside the different factions (sometimes shifting allegiances, it has to be added). As of 1976 and through until 2005, the Syrian military occupied Lebanon, controlling thereby the country's leadership and parliamentary composition causing many Christians to flee (around a quarter of a million). As the Civil War drew to a close, a Peace Treaty between Lebanon and Syria was signed in 1991, proclaiming that Lebanon could not threaten Syria's security whilst Syria would protect Lebanon from external threats (effectively, this meant the Syrian military remained and were still in control). After 30 years of this occupied status, Lebanese civil resistance culminated in the Cedar Revolution (2005), forcing the Syrian military, finally, to withdraw.

Currently, Lebanon is again a host country to a new wave of refugees arriving from close-by Middle-Eastern countries in thrall to civil conflicts or wars of their own – amongst them some 1.5m Syrians (the irony will escape none of you). A further remarkable feature of the present refugee crisis is that a third of the 26m refugees worldwide are hosted by the world's poorest countries, two of which are Lebanon and Jordan. Furthermore, Lebanon has the highest count, world-wide, of refugees per thousand (160; Jordan stands at 72; next is Turkey with 45).[53]

In this migration mess the effects of climate change also have their role to play (the other transgression of the ecology ceiling on our Doughnut above). Even if the Syrian refugees are present in Lebanon because of the devastating Civil War in their own country, to a lesser degree the effects of successive droughts (2006–9) in the rural North-East (known as the Fertile Crescent) coupled with changes in economic practices by President Assad (cutting agricultural subsidies, privatising state farms, etc.) affected the migratory pattern within

---

53 www.amnesty.org/en/what-we-do/refugees-asylum-seekers-and-migrants/global-refugee-crisis-statistics-and-facts/ (accessed 17 November 2019).

Syria. Around 60,000 families from that region migrated to urban areas, thus adding pressure on the already under-resourced cities themselves.[54] By the time of the Arab Spring uprisings of 2010, all manner of social and economic injustices were biting into the fabric of Syria's society, leading to their own anti-government movement in 2011 which was the trigger for the ongoing Civil War.

Again in terms of climate change, given Rahil's plight within the context of *Capharnaüm*, we should consider Ethiopia, which is one of the world's most drought-prone countries. High temperatures along with irregular rainfall leading to soil erosion have affected agricultural production. Combined with an increasing population and conflicts, food insecurity runs high. All of which explains the predominantly rural labour migration towards the Middle-East. And in Lebanon, Ethiopians constitute the largest migrant group of workers. Whilst many are there illegally, even if the migrant is legally present in the host nation, typically that person is indentured to the government's Kafala Sponsorship System. The system requires all unskilled labourers to have an in-country sponsor, usually their employer, who is responsible for their visa and legal status. But many employers take away passports and abuse their workers with little chance of legal repercussions.[55] The scene in the film where Rahil's fun-fair friends try to pose as her employing sponsors now makes greater sense – since neither Rahil nor they have possession of her papers (legal or illegal), nothing can be done to help her and she cannot escape the tentacles of the exploitative Aspro.

Lastly, on this issue of migration, half of the world's refugees are children.[56] And this brings us back to the central thrust of *Capharnaüm*. Labaki has spoken of her drive to make this film as a protest against the normalisation of child poverty. 'The begging of hundreds of children in the streets (of Beirut) has become the

---

54 See Jan Selby, Omar S. Dahi, Christiane Frölich, Mike Hulme, 'Climate change and Syrian civil war revisited,' www.sciencedirect.com/science/article/pii/S0962629816301822 (accessed 14 November 2019).
55 See www.reuters.com/article/us-lebanon-migrants-irregular/trapped-by-the-system-ethiopian-workers-in-lebanon-see-no-freedom-idUSKCN1FZ195 (accessed 12 November 2019). There are also reports of high suicide rates amongst domestic foreign workers. The government is generally perceived as being lax in protecting the human rights of migrant workers. See https://en.wikipedia.org/wiki/Kafala_system (accessed 14 November 2019).
56 See www.amnesty.org/en/what-we-do/refugees-asylum-seekers-and-migrants/global-refugee-crisis-statistics-and-facts/ (accessed 15 November 2019).

new normal,' she declared in an interview, adding, 'how do we allow for such injustice to happen to the most fragile human beings in society?'[57] The terribly damaged state of Lebanon, in truth on the brink of bankruptcy, is a part answer to the «how». Yet this is difficult to conciliate with the massive display of wealth alongside the coast-line of Beirut and the rebuilding of the city centre, for example; a bubble in which the nation's top 2% can live in luxury and amuse themselves whilst remaining 'oblivious' to the slum conditions surrounding them.[58]

As Labaki comments, 'by suing his parents, (Zain) is also suing a system, a whole society that is not allowing him to have his basic rights.'[59] Human rights that would include education, decent housing, water and food, and ultimately a chance to work for a decent income – all of which are in shortfall, as you can note from the red list of critical human deprivation within the Doughnut (Figure 4.4). However, Lebanon is ranked as 129th out of 141 countries in terms of income equality. The figures for children in education only apply to those children who are registered at birth – and as we know Zain and his siblings are not registered. Even so, only 68% of the official total sector is in secondary education.[60] Lebanon's economy has a national debt of 150% of GDP, and central bank reserves have plunged 30% over the past year.[61]

Whilst much of this desolate state can be imputed to the effects of the 30 years of both Civil War and Syrian occupation, the fact remains that poverty – at 25% of the Lebanese population, and at 33% once the official figures for the refugee and migrant populations are included – is rampant. Moreover, 20% of the population live in

---

57 www.npr.org/2018/12/16/676553757/in-capernaum-the-chaos-of-lebanon-from-a-homeless-child-s-perspective (accessed 11 October 2019).
58 See two very interesting articles on this issue of rebuild and wealth: www.theguardian.com/cities/2015/jan/22/beirut-lebanon-glitzy-downtown-redevelopment-gucci-prada and www.nytimes.com/2018/02/26/lens/a-glimpse-into-the-world-of-lebanons-1-percent.html (accessed 18 November 2019). Moreover, in Diop's film *Atlantique*, a similar situation arises with the luxury tower being constructed on the beachfront in Dakar whilst the poor, including the building site labourers, have nothing.
59 www.npr.org/2018/12/16/676553757/in-capernaum-the-chaos-of-lebanon-from-a-homeless-child-s-perspective (accessed 11 October 2019).
60 www.undp.org/content/dam/lebanon/docs/Governance/Publications/Assessing%20Labor%20Income%20Inequality%20in%20Lebanon's%20Private%20Sector.pdf (accessed 18 November 2019).
61 www.theguardian.com/world/2019/oct/20/lebanons-mass-revolt-against-corruption-and-poverty-continues (accessed 18 November 2019).

squalor (with poor sanitation and either unclean drinking water or no running water at all).[62] Meantime, in the private sector, the top 2% capture a share of the income as high as that of the bottom 60%.[63] And of course with 50% of the Lebanese labour force (to say nothing of the 'illegals') working in the informal sector, to a large extent in lower (poor) income brackets, the actual level of poverty and inequality is hard to estimate in real figures.[64] Demonstrators are once again out on the streets of the major Lebanese cities revolting against corruption and poverty in the largest civic protests since the Cedar Revolution, this time, rightfully claiming that governmental officials are 'preventing badly needed reforms that would cut into pockets of the ruling class, and are instead trying to recoup state revenues by taxing the poor.'[65]

Capharnaüm stands as a documentary memory to this state of affairs, one which is not just the case of Lebanon, but the case of world injustice. Everything was shot in the raw-real, as Labaki tells us:

> in the prisons with the prisoners – these are real prisons. The apartments are real apartments – the drawings on the walls are drawings made by kids who lived in these apartments. The rule was to intervene as least as possible in order to be able to tell the truth.[66]

The film makes a powerful statement against child poverty and the causes and abuses of migration. And, whilst Zain is undoubtedly the central protagonist of this film, Rahil/Tigest remains – for me at least – a symbol of the path to follow. Both her names tell us so: Rahil meaning path-guider and her adopted name, Tigest, patience. More than this, Tigest also shows the «how» of this path. People of this

---

62 https://borgenproject.org/top-10-crucial-to-know-facts-about-poverty-in-lebanon/ (accessed 18 November 2019).
63 www.undp.org/content/dam/lebanon/docs/Governance/Publications/Assessing%20Labor%20Income%20Inequality%20in%20Lebanon's%20Private%20Sector.pdf (accessed 18 November 2019).
64 www.undp.org/content/dam/lebanon/docs/Governance/Publications/Assessing%20Labor%20Income%20Inequality%20in%20Lebanon's%20Private%20Sector.pdf (accessed 18November 2019).
65 www.theguardian.com/world/2019/oct/20/lebanons-mass-revolt-against-corruption-and-poverty-continues (accessed 18 November 2019).
66 Labaki quoted in www.npr.org/2018/12/16/676553757/in-capernaum-the-chaos-of-lebanon-from-a-homeless-child-s-perspective (accessed 11 October 2019).

name have a deep inner desire to save humanity and give to others by sharing money, knowledge, experiences, and creative artistic ability.

Is there more to be said?

Figure 4.5 The Lebanese Cedar, a symbol of hope and regeneration.[67]

---

67 Photo accessed from https://pixabay.com/photos/tree-cedar-ancient-wood-3663738/ (accessed 17 December 2019).

Bibliography

Acemoglu, D. & Robinson, J. A. (2013) *Why Nations Fail: The Origins of Power, Prosperity and Poverty*, London, Profile Books Ltd.

Baudrillard, J. (1996) *The System of Objects*, Transl. James Benedict, London and New York, Verso (originally printed 1968, *Le Système des objets*, Paris, Gallimard).

Crenshaw, K. (1991) 'Mapping the Origins: Intersectionality, Identity Politics, and Violence against Women of Color', *Stanford Law Review*, Vol. 43, No. 6, pp. 1241–99.

Deleuze, G. (1985) *Cinéma 2: L'Image-temps*, Paris, Éditions de Minuit.

Demos, T. J. (2017) *Against the Anthropocene: Visual Culture and Environment Today*, Berlin, Sternberg Press.

DuBois Shaw, G. (2004) *Seeing the Unspeakable: The Art of Kara Walker*, Durham, NC and London, Duke University Press.

Eddo-Lodge, R. (2017) *Why I'm No Longer Talking to White People about Race*, London, Bloomsbury Circus.

Fukuyama, Y. F. (2002) *Our Posthuman Future: Consequences of the Biotechnology Revolution*, New York, Farrar, Strauss and Giroux.

Gagnier, R. (2018) *Literatures of Liberalization: Global Circulation and the Long Nineteenth Century*, Cham, Palgrave Macmillan.

Kilpi, H. (2013) 'The Landscape's Lie: Class, Economy, and Ecology in *Hotel Rwanda*', In: Paula Willoquet-Maricondi (ed), *Framing the World in Ecocriticism*, Charlotteville and London, University of Virginia Press (135–53).

Klein, N. (2007) *The Shock Doctrine: The Rise of Disaster Capitalism*, Knopf Canada.

Marks, Laura, U. (2000) *The Skin of the Film: International Cinema, Embodiment, and the Senses*, Durham, NC and London, Duke University Press.

Raworth, K. (2017) *Doughnut Economics: Seven Ways to Think like a 21-st Century Economist*, London, Penguin Random House.

Rodowick, D. N. (1997) *Gilles Deleuze's Time Machine*, Durham, NC and London, Duke University Press.

Varoufakis, Y. (2017) *Talking to My Daughter about the Economy: A Brief History of Capitalism*, London, The Bodley Head.

Bibliography

Annotated bibliography: film and ecocriticism

Bozak, N. (2011) *The Cinematic Footprint: Lights, Camera, Natural Resource*, New Brunswick, NJ and London, Rutgers University Press.

Hjort, M. (ed) (2012) *Film and Risk*, Detroit, MI, Wayne University Press.

Ingram, D. (2000) *Green Screen: Environmentalism and Hollywood Cinema*, Exeter, University of Exeter Press.

Kibbey, A. (2005) *Theory of the Image: Capitalism, Contemporary Film and Women*, Bloomington and Indianapolis, Indiana University Press.

Murray, R. L. & Heumann, J. K. (2014) *Film and Everyday Eco-Disasters*, Lincoln and London, University of Nebraska Press.

Rust, S., Monani, S. & Cubitt, S. (2013) *Ecocinema: Theory and Practice*, London and New York, Routledge.

Willoquet-Maricondi, P. (2010) *Framing the World in Ecocriticism*, Charlottesville and London, University of Virginia Press.

Index

Note: Page numbers followed by "n" denote footnotes.

Adetiba, Kemi 60
Afolayan, Kunle 60
The Age of Stupid (2009) 62
Akinmolayan, Niyi 60
Anyaene, Chineze 60
Après Demain (2018) 61
Armstrong, Franny 23, 62
Atlantique (2019) 76, 109n58

Bandele, Biyi 60
Baudrillard, Jean 17, 19, 27, 29
Behind the Beautiful Forevers (2014) 24
Berg, Peter 18
Bigelow, Kathryn 64
Blade Runner (1982) 64

Capharnaüm/Chaos (2018) 75, 76, 81, 99–111
Cathy Come Home (1966) 16
The China Syndrome (1979) 19
Curtiz, Michael 33

The Day After Tomorrow (2004) 19
The Day the Earth Stood Still (1951) 19
Deepwater Horizon (2016) 18–9
Demain/Tomorrow (2015) 61, 62
Demos, T. J. 17–20, 23, 48
Dion, Cyril 61
Diop, Mati 76, 109n58
Doughnut Economics/Doughnut Theory 2, 3, 4, 13–15, 20, 21, 33–6, 39, 40, 44–5, 47–58, 62, 64, 67–71, 74–5, 81, 83, 85, 91–2, 95–7, 105–9
Drowned Out (2002) 23

Erin Brockovich (2000) 46, 62, 64–9, 70, 71–2, 74–5

The Figurine (2009) 60
For Sama (2019) 99
Fuocamare/Fire at Sea (2016) 2

George, Terry 81, 92–3, 95, 99
Giant (1956) 17, 33, 36, 40–5, 63
Les Glaneurs et la glaneuse/The Gleaners and I (2000) 63

Half a Yellow Sun (2013) 60
Hare, David 24
Heisler, Stuart 33
Hidden Figures (2016) 8–9, 62
Hotel Rwanda (2004) 75, 81–9, 100
Human Flow (2017) 2, 15

I Daniel Blake (2016) 16
Ijé/The Journey (2010) 60

Al-Kateab, Waad 99
Kramer, Stanley 18
Kubrick, Stanley 18

Labaki, Nadine 75, 99, 108–9, 110
Laurent, Mélanie 61

Index

Loach, Ken 16, 30, 61, 62–3
Lucas, George 64

McDonagh, Martin 62
McQueen, Steve 76
Mad Max Fury Road (2015) 19
The Matrix (1999) 64
Melfi, Theodore 8
Mildred Pierce (1945) 17, 33–6, 39, 44, 45, 63

On the Beach (1959) 18

Paddington Bear series 62
Pasolini, Pier Paolo 71
Pilger, John 20
Planet of the Apes (1968) 19

Raworth, Kate 2, 3, 4, 13–16, 20, 21, 35, 39, 43–4, 47–59, 60–4, 69, 71, 75, 81, 86, 96
Redgrave, Vanessa 2
Rosi, Gianfranco 2

Salò, or the 120 Days of Sodom (1975) 71
Sans toit ni loi (1984) 60
Scorsese, Martin 47
Scott, Ridley 64
Sea Sorrow (2017) 2

Soderbergh, Steven 47
Sorry We Missed You (2019) 16, 63
Star Wars (1977) 64
Stevens, George 33
Stone, Oliver 47
Strange Days (1995) 64

Tehran Taxi (2015) 62
3 Billboards Outside Ebbing, Missouri (2017) 62
Tulsa (1949) 33, 36, 37–40, 44, 45, 63
Twelve Years a Slave (2014) 76
2001: A Space Odyssey (1968) 18

Utopia (2013) 20

Varda, Agnès 61, 63
Von Einsiedel, Orlando 62

Wachowski, Lana 64
Wachowski, Lilly 64
Wall Street (1987) 46, 47, 64, 69–75
The Wedding Party (2016) 60
Weiwei, Ai 2, 15
White Helmets (2016) 62
The Wind that Shakes the Barley (2006) 63
Wise, Robert 19
Wolf of Wall Street (2013) 46, 47, 64, 69–75

For Product Safety Concerns and Information please contact our EU representative GPSR@taylorandfrancis.com
Taylor & Francis Verlag GmbH, Kaufingerstraße 24, 80331 München, Germany

www.ingramcontent.com/pod-product-compliance
Lightning Source LLC
Chambersburg PA
CBHW051754230426
43670CB00012B/2284